T0135615

Flexible Semi- and Non-Parametric Modelling and Prognosis for Discrete Outcomes

Inaugural-Dissertation
zur Erlangung des Grades Doctor oeconomiae publicae (Dr. oec. publ.)
an der Ludwig-Maximilians-Universität München

vorgelegt von

Harald Binder

Jahr: 2005

Referent: Prof. Dr. G. Tutz
Korreferent: Prof. Dr. L. Fahrmeir

Promotionsabschlussberatung: 8.2.2006

Bibliografische Information Der Deutschen Bibliothek

Die Deutsche Bibliothek verzeichnet diese Publikation in der Deutschen
Nationalbibliografie; detaillierte bibliografische Daten sind im Internet über
http://dnb.ddb.de abrufbar.

ISBN 3-8325-1172-7

Logos Verlag Berlin
Comeniushof, Gubener Str. 47,
10243 Berlin
Tel.: +49 030 42 85 10 90
Fax: +49 030 42 85 10 92
INTERNET: http://www.logos-verlag.de

Vorwort

Statistische Forschung findet oft in einem Spannungsfeld zwischen mathematischen Grundlagen und konkreter Anwendung statt. Auch die vorliegende Arbeit ist exemplarisch dafür: Sie entstand während meiner Tätigkeit als wissenschaftlicher Mitarbeiter für Methoden und Statistik an der Klinik und Poliklinik für Psychiatrie, Psychosomatik und Psychotherapie der Universität Regensburg – und parallel gedieh, mehr oder minder stetig, meine Statistik-Promotion an der Ludwig-Maximilians-Universität München. An dieser Stelle danke ich deshalb zuallererst meinem Doktorvater Prof. Gerhard Tutz. Ohne seine inspirierenden Ideen, seine kritischen Nachfragen und seinen optimistischen Ansporn würde diese Arbeit sicher in deutlich geringerem Maße Kriterien fundierter statistischer Forschung genügen. Für ein forschungsfreundliches Klima, in dem neben dem Datenanalyse-Alltag noch Raum für die Entwicklung neuer Methoden blieb, gebührt mein Dank Herrn Prof. Helmfried Klein und Herrn Prof. Clemens Cording. Für seine Wegweisung auf meinem Weg von der inhaltlich-psychologischen Forschung zur Fokussierung auf Methoden bin ich Herrn Prof. Bill Batchelder herzlich dankbar. Auf die vergangenen Jahre zurückblickend, möchte ich vor allem auch jene gesunde Mischung nicht missen, die sich aus "forschungsfreien" Interaktionen und vielfältigsten Mittagspausen-Diskussion mit meinen Kollegen am Klinikum ergab. Dabei denke ich auch an meinen Zimmerkollegen Winfried Barta, der sich allen halbfertigen, unausgereiften oder minder fundierten Gedanken beharrlich widersetzte, und dessen trockene Kommentare so manchen Tag in anderes Licht setzten. Meiner Familie möchte ich schließlich dafür danken, daß sie es mir ermöglichte, einen Berufsweg nach meinen Neigungen zu wählen, und daß sie diesen Weg nach Kräften unterstützte. Ganz inniger Dank gilt Claudia. Ohne ihre liebevolle motivationale, logistische und gedankliche Unterstützung hätte ich die vergangenen Jahre längst nicht so gut überstanden.

Abstract

This thesis presents new models and estimation techniques based on the generalized linear model framework. In Chapter 1 the GAMBoost procedure for estimation of generalized additive models is developed. Based on boosting and gradient descent in function space it generalizes the notion of repeated fitting of residuals to exponential family responses. By using a flexible number of updates for each covariate the selection of smoothing parameters is reduced to the selection of the number of boosting steps. The latter is chosen based on approximate effective degrees of freedom. The resulting procedure shows good performance for a wide variety of examples with binary and Poisson response data. A considerable advantage compared to other procedures is found for a large number of covariates and a low level of information. An application to real data is presented. In Chapter 2 a flexible model for discrete time survival data is developed that allows for non-linear covariate effects that vary over time. For estimation an iterative two-step procedure based on Fisher scoring is given. A simulation study with various levels of complexity underlying the data compares the performance of adequate models to the performance of models that are too restrictive, too flexible or that provide the wrong kind of flexibility. It is shown that effective degrees of freedom work well as a basis for selection of smoothing parameters as well as for model selection. An example with real data is given. In Chapter 3 a technique for classification with binary response data is developed based on logistic regression. It uses local models with local selection of predictors for reduction of complexity and penalized estimation for numerical stability. Standard simulated data examples are used to evaluate components of the algorithm such as the kernel for local weight calculation, and to compare the performance to other procedures. It is found that the procedure is competitive for a wide range of examples and that selection of predictors is crucial for local quadratic models while being regulated rather well for local linear models. Good performance can also be seen for real data examples.

Contents

Introduction   1

1   Generalized additive models by additive boosting   3

   1.1   Generalized linear and additive models 4

       1.1.1   Framework 4

       1.1.2   Estimation 7

   1.2   Generalized additive model boosting 10

       1.2.1   Boosting and gradient descent 10

       1.2.2   Generalized additive models by repeated fitting of residuals   11

       1.2.3   Specific learners 14

   1.3   Model characteristics 18

       1.3.1   Selection of parameters and model complexity 18

       1.3.2   Confidence bands 23

   1.4   Comparison to alternative fitting methods 25

       1.4.1   GAMBoost and LogitBoost 25

       1.4.2   Empirical comparison 25

1.5   Real data 36

1.6   Discussion 38

2 Flexible modelling of discrete failure time   39

   2.1   Theory 40

       2.1.1   Framework for survival models 40

       2.1.2   Flexible models for discrete time 41

   2.2   Simulation study 50

       2.2.1   Parametric and non-linear effects without time-variation . 55

       2.2.2   Parametric and non-linear effects with time-variation . . . 58

   2.3   Real data 64

   2.4   Discussion 67

3 Localized logistic regression   69

   3.1   Theory 71

       3.1.1   Localized logistic regression 72

       3.1.2   Local reduction of dimensions by selection of predictors . . 74

       3.1.3   Computational optimization 75

       3.1.4   Algorithm and parameter selection 77

       3.1.5   Relevance of variables 78

   3.2   Simulation study 78

       3.2.1   Example data and results 80

       3.2.2   Summary of simulation results 92

3.3 Real data 95

3.4 Discussion 98

Introduction

Regression models form the core of classical statistics and are the basis of much of statistical practice. Their development dates back to the early 19th century with contributions from Legendre and Gauss and there is a vast amount of research extending the regression model framework in various ways. The development of regression models for discrete response data originates in the early 20th century with models that relate proportions of outcomes to experimental conditions (e.g. Bliss, 1935; Fisher, 1935). Later Nelder and Wedderburn (1972) introduced the generalized linear model framework which allows for unified treatment of regression models with an exponential family response. One important extension of this framework was to provide more flexibility with respect to the shape of covariate influence. Generalized additive models, as introduced by Hastie and Tibshirani (1986), allow for arbitrary unknown functions for covariate influence. For example in the backfitting algorithm any kind of scatterplot smoothers can be used to estimate covariate influence.

This work introduces new models and estimation techniques for discrete outcomes, based on the generalized linear model framework. The focus is on providing additional flexibility, in the lines of generalized additive models, by using non-parametric techniques for modelling and prognosis.

In Chapter 1 of this thesis a new estimation technique for generalized additive models is developed. The use of current techniques is problematic due to numerical problems when there is a large number of covariates. In addition the choice of a smoothing parameter for each of the covariates results in a high-dimensional optimization problem, which is very computationally demanding. A stepwise approach is introduced, that is based on boosting ideas (Friedman et al., 2000; Bühlmann and Yu, 2003) and is applicable to a large number of potential predictors and automatically determines the amount of smoothness required for each covariate. This is demonstrated in an extensive simulation study and the algorithm is applied to a real data example.

1

Chapter 2 considers the special structure of discrete time survival models. A model framework is introduced that does not only allow for non-linear influence of covariates, but also for variation of the covariate effect in the course of time. Such models for example are useful for modelling the effect of covariates that are recorded only once, because the influence of such initial measurements on the response might diminish as time progresses. Providing such flexibility leads to the danger of overfitting. Therefore the issue of model selection is investigated in a simulation study. A real data example is used to demonstrate the application of the model framework.

In Chapter 3 the focus is on prediction performance for binary response data. Local logistic regression, as discussed by Loader (1999), is used as a basis to develop a new classification algorithm. For application to high-dimensional problems a robust version with an additional variable selection step is introduced. This procedure is shown to be competitive with up-to-date classification procedures on a number of standard simulated and real data examples.

Parts of the present thesis are also found in the papers "Tutz, G. and Binder, H. (2004). Generalized additive modelling with implicit variable selection by likelihood based boosting. Discussion Paper 401, SFB 386, Ludwig-Maximilians-University Munich.", "Tutz, G. and Binder, H. (2004). Flexible modelling of discrete failure time including time-varying smooth effects. *Statistics in Medicine*, 23(15): 2445-2461.", "Binder, H. and Tutz, G. (2004). Localized logistic classification with variable selection. In Antoch, J., editor, *COMPSTAT 2004*, 705-712, Heidelberg. Physica-Verlag.", "Tutz, G. and Binder, H. (2005). Localized classification. *Statistics and Computing*, 15(3):155-166.".

Chapter 1

Generalized additive models by additive boosting

The starting point for this chapter is a set of data, containing n observations (y_i, x_i), $i = 1, \ldots, n$ where $x_i' = (x_{i1}, \ldots, x_{ip})$ is a vector of p covariates and y_i is the observed response. What is wanted is a model that allows to explore the relation of the covariates x_i to the response y_i. It is assumed that the distribution of y_i belongs to a simple exponential family and specifically the focus is on discrete response data. For flexible modelling generalized additive models (Hastie and Tibshirani, 1986; Buja et al., 1989; Hastie and Tibshirani, 1990) are chosen which put no specific restrictions on the shape of the influence of every single covariate.

In Section 1.1 the generalized linear model framework, generalized additive models and current methods of estimation are reviewed. In Section 1.2 a new estimation technique for generalized additive models is introduced that allows for a large number of covariates and automatically detects the amount of flexibility that is required for each covariate. In Section 1.3 pointwise confidence bands for the resulting estimates and approximate degrees of freedom are derived. Based on the latter the selection of model parameters is discussed. The performance of the new method is compared to that of classical fitting techniques in Section 1.4 and an application to real data is shown in Section 1.5. In Section 1.6 the benefits of the procedure are discussed and guidelines for its application are given.

1.1 Generalized linear and additive models

1.1.1 Framework

In classical linear regression the relation of the covariates $x_{ij}, j = 1, \ldots, p$ to the response y_i is described by the model

$$y_i = \beta_0 + \sum_{j=1}^{p} \beta_j x_{ij} + \epsilon_i \quad i = 1, \ldots, n, \tag{1.1}$$

using parameters $\beta_j, j = 0, \ldots, p$ and an error term ϵ_i. The latter accounts for all variation of the response not accounted for by the covariates. It is assumed that ϵ_i is independently identically distributed with

$$\epsilon_i \sim N(0, \sigma^2).$$

Using a design vector z_i built from the covariates by $z_i' = (1, x_{i1}, \ldots, x_{ip})$ and the parameter vector $\beta' = (\beta_0, \ldots, \beta_p)$ the model can be expressed as

$$y_i = z_i'\beta + \epsilon_i \quad i = 1, \ldots, n.$$

Given the data, estimates for the parameter vector β are obtained by least-squares estimation.

The generalized linear model framework (GLM) introduced by Nelder and Wedderburn (1972) generalizes the linear regression model to a wide range of response types (including metric, binary and counting response data). It distinguishes between the *structural part*

$$\mu_i = h(\eta_i) = h(z_i'\beta) \tag{1.2}$$

where h is a known response function and the *distributional assumption* stating that the distribution of y_i belongs to a simple exponential family with expectation

$$E(y_i|x_i) = \mu_i$$

and density

$$f(y_i|\theta_i, \phi, \omega_i) = \exp\left(\frac{y_i\theta_i - b(\theta_i)}{\phi}\omega_i + c(y_i, \phi, \omega_i)\right)$$

where θ_i is the so-called natural parameter, ϕ is a scale or dispersion parameter, b and c are functions corresponding to the specific type of exponential family and

4

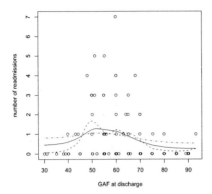

Figure 1.1: Number of readmissions within three years after discharge for patients with different levels of functioning (indicated by the GAF score) at discharge (with some jitter added). Penalized B-spline estimates with various amounts of smoothing are indicated by solid, broken and dash-dotted lines.

ω_i is a weight for the observation. The natural parameter is a function of the mean, i.e., $\theta_i = \theta(\mu_i)$, which is uniquely determined by $\mu = \partial b(\theta)/\partial \theta$. The mean structure also implies a certain variance structure of the form

$$\mathrm{var}(y_i|x_i) = \phi v(\mu_i)/\omega_i,$$

with $v(\mu) = \partial^2 b(\theta)/\partial \theta^2$. For more details see McCullagh and Nelder (1989) or for example Fahrmeir and Tutz (2001) for a more recent treatment.

When y_i given x_i is taken to be from a normal distribution and h is the identity, i.e., $\mu_i = \eta_i$, then the classical linear model (1.1) is recovered. The logit model or logistic regression, a standard tool for binary response data, is obtained when a binomial distribution $y_i \sim B(1, \mu_i)$ is assumed with

$$\mu_i = h(\eta_i) = \frac{\exp(\eta_i)}{1 + \exp(\eta_i)}.$$

In this chapter in addition the focus is also on models for counting responses. For the latter it is assumed that y_i is from a Poisson distribution $y_i \sim Poisson(\mu_i)$ with

$$\mu_i = h(\eta_i) = \exp(\eta_i).$$

5

For an example where the linear predictor used in (1.2) is problematic consider the data in Figure 1.1. It shows the number of readmissions within three years after discharge for 94 patients of a psychiatric hospital plotted against a score for the patients' level of functioning (GAF score) at discharge (for more details see Section 1.5). In principle a Poisson response model might be adequate and therefore the generalized linear model framework could be applied. However, it is obvious that a linear predictor of the form "$\eta_i = \beta_0 + \beta_1 \cdot \mathrm{GAF}$ at discharge" is inadequate, because the maximum number of readmissions occurs at a medium level of the covariate.

When there are no specific assumptions about the shape of the covariate influence, the most attractive approach is to allow for arbitrary smooth functions of covariates. There is a vast literature on scatterplot smoothers that attempt to fit such smooth functions. For an overview of methods in the context of generalized linear models see for example Hastie and Tibshirani (1990) or Green and Silverman (1994). Most of these methods provide a single regularization parameter that determines the smoothness of the fitted function. The curves in Figure 1.1 indicate three fits with different smoothing parameters. While the solid curve seems to be adequate, the smoothness constraint put on the dash-dotted curve might be too strong. The broken curve exhibits several features not seen in the other curves (e.g. two local maxima) that might not reflect true underlying structure but just noise in the data. The latter phenomenon is also called overfitting. To prevent this from happening while not missing important features the smoothing parameter has to be chosen thoroughly.

Hastie and Tibshirani (1986) introduced generalized additive models, an extension of generalized linear models, by allowing allowing for arbitrary functions $f_j, j = 1, \ldots, p$ for the influence of each covariate. The predictor then has the form

$$\eta_i = \beta_0 + \sum_{j=1}^{p} f_j(x_{ij}). \qquad (1.3)$$

A feature that is preserved from generalized linear models is that the covariates are assumed to be additive in their effect on the predictor. Interactions have to be included explicitly by adding functions of two (or more) covariates, e.g. $f_{jk}(x_{ij}, x_{ik})$, to the predictor. Besides computational advantages this restriction aids interpretation of the model fits, because often interactions of high dimension are difficult to understand.

6

1.1.2   Estimation

An estimate for the parameter vector β of a generalized linear model (1.2) is obtained by maximizing the likelihood given the observations. In the following this is sketched shortly. For a comprehensive treatment see McCullagh and Nelder (1989) or for example Fahrmeir and Tutz (2001).

The log-likelihood contribution of observation i is (up to an additive constant)

$$
\begin{aligned}
l_i(\theta_i) &= \log f(y_i|\theta_i, \phi, \omega_i) = \frac{y_i \theta_i - b(\theta_i)}{\phi} \omega_i = \frac{y_i \theta(\mu_i) - b(\theta(\mu_i))}{\phi} \omega_i \\
&= l_i(\mu_i) = l_i(h(z_i'\beta)) = l_i(\beta).
\end{aligned}
$$

The log-likelihood of the sample is the sum of the individual contributions

$$
l(\beta) = \sum_i l_i(\beta). \tag{1.4}
$$

The maximum likelihood estimate $\hat{\beta}$ of β is obtained by solving to the likelihood equation

$$
s(\hat{\beta}) = 0
$$

where $s(\beta)$ is the score function

$$
s(\beta) = \frac{\partial l}{\partial \beta}.
$$

This can be done by iterative use of Fisher scoring steps

$$
\hat{\beta}^{(k+1)} = \hat{\beta}^{(k)} + F^{-1}(\hat{\beta}^{(k)}) s(\hat{\beta}^{(k)}), \quad k = 0, 1, 2, \ldots
$$

where $F(\beta)$ is the (observed) Fisher information matrix

$$
F(\beta) = \frac{\partial^2 l(\beta)}{\partial \beta \partial \beta'}.
$$

Each iteration can be interpreted as a weighted least squares fit with working response

$$
\tilde{y}_i^{(k)} = \hat{\eta}_i^{(k)} + \frac{(y_i - h(\hat{\eta}_i^{(k)}))}{\partial h(\hat{\eta}_i^{(k)})/\partial \eta} \tag{1.5}
$$

and weights

$$
w_i = \frac{\partial h(\hat{\eta}_i^{(k)})}{\partial \eta} \hat{\sigma}^2 \tag{1.6}
$$

7

where $\hat{\eta}_i^{(k)}$ is evaluated at the estimate $\hat{\beta}^{(k)}$ from the previous step and $\hat{\sigma}^2 = \text{var}_{\hat{\beta}^{(k)}}(y_i)$. This is useful for adapting methods developed in the least-squares setting.

The method proposed by Hastie and Tibshirani (1990) for fitting additive models (i.e., with Gaussian response) is *backfitting*. The basic principle of this procedure is to fit the component for each covariate by fitting the partial residuals from the other components in an alternating fashion. Using vector notation, with y being the centered response vector and $f_j^{(m)} = (f_j(x_{1j}), \ldots, f_j(x_{nj}))', j = 1, \ldots, p$ the fitted smooth components after step m, the update in step $m+1$ can be written as

$$\hat{f}_j^{(m+1)} = S_j \left(y - \sum_{s<j} \hat{f}_s^{(m+1)} - \sum_{s>j} \hat{f}_s^{(m)} \right) \quad j = 1, \ldots, p,$$

where S_j represents the smoother to be used for component j. S_j typically will be a smoother matrix, but in principle any scatterplot smoother can be used.

For non-Gaussian response backfitting is applied to the working response (1.5) with weights (1.6). The working response and weights are recalculated and backfitting is applied again, until the estimates do not change anymore. This procedure is called *local scoring* (Hastie and Tibshirani, 1990, p. 141).

There are several alternatives and extension to backfitting available. Hastie and Tibshirani (2000) apply the Gibbs sampler to additive models resulting in a Bayesian backfitting procedure. This allows for example to account for repeated measurements by including a random intercept term. Fahrmeir and Lang (2001) give a very flexible Bayesian approach for generalized additive models based on the Metropolis-Hastings algorithm.

In the following the focus is on an approach that, in contrast to backfitting, estimates all smooth functions simultaneously (see e.g. Marx and Eilers, 1998). This alternative is available when the smoothers can be represented by a basis expansion

$$f_j(x_{ij}) = \sum_{l=1}^{M} \beta_{jl} B_l^{(j)}(x_{ij})$$

with known basis functions $B_l^{(j)}(x), l = 1, \ldots, M$, which preferably should be local. For example Eilers and Marx (1996), whose P-spline approach will be adapted in the following, use a B-spline basis with a large number of equally spaced knots. For a comprehensive treatment of B-splines see de Boor (1978).

When a design vector

$$z_i' = (1, B_1^{(1)}(x_{i1}), \ldots, B_m^{(1)}(x_{i1}), \ldots, B_1^{(p)}(x_{ip}), \ldots, B_m^{(p)}(x_{ip}))$$

is formed, the standard generalized linear model framework applies and estimation can be based on Fisher scoring. For algorithms that also include the selection of smoothing parameters see Wood (2000, 2004). Complexity of the model to be estimated can be kept at a reasonable amount either by using only few basis functions or by putting smoothness restrictions on the parameter estimates. When only few basis functions are used B-splines for example no longer are as local in their effect as might be wanted. Therefore the latter approach of using a large number of basis functions (say 30) and restricted parameter estimation is employed. This is implemented by maximization of a penalized version of the log-likelihood (1.4)

$$l_p(\beta) = l(\beta) + \beta' P \beta \tag{1.7}$$

with parameter vector $\beta' = (\beta_0, \beta_{11}, \ldots, \beta_{1M}, \ldots, \beta_{p1}, \ldots, \beta_{pM})$. The penalty matrix P is block-diagonal with

$$P = \begin{pmatrix} 0 & & & \\ & \lambda_1 P_1 & & \\ & & \ddots & \\ & & & \lambda_p P_p \end{pmatrix}$$

where $P_j, j = 1, \ldots, p$ are penalty matrices for the individual components and the penalty parameters $\lambda_j, j = 1, \ldots, p$ determine the amount of smoothing to be used for each covariate. In the following B-spline basis functions are used. Eilers and Marx (1996) demonstrate how a penalty matrix P_j for these has to be structured to obtain various types of smoothness for the estimates. When the derivatives of the estimated functions are not of interest, typically second order penalization is applied. Therefore individual penalty matrices of the form

$$P_j = D_j' D_j \quad \text{with} \quad D_j = \begin{pmatrix} -1 & 1 & & \\ & -1 & 1 & \\ & & \ddots & \\ & & & -1 & 1 \end{pmatrix}, \quad j = 1, \ldots, p$$

are used, which result in a penalization of differences $\sum_l (\beta_{j,l+1} - \beta_{jl})^2$. Note that for reasons of uniqueness centered estimates with the restriction $\sum_l \beta_{jl} = 0$ are used.

The Fisher scoring steps for estimation of β by maximization of (1.7) take the form

$$\hat{\beta}^{(k+1)} = \hat{\beta}^{(k)} + F_p(\hat{\beta}^{(k)})^{-1} s_p(\hat{\beta}^{(k)}), \quad k = 0, 1, 2, \ldots \tag{1.8}$$

with the penalized Fisher matrix $F_p(\beta) = F(\beta) + 2P$ and penalized score function $s_p(\beta) = s(\beta) - 2P\beta$.

One problem that for example for backfitting is hidden in the notation is the selection of smoothing parameters. As for each covariate a separate smoothing parameter should be allowed for, a general approach is to evaluate several combinations based on a criterion such as prediction performance by cross-validation. When there is a large number of covariates the problem of selecting smoothing parameters becomes more severe, because the optimal combination effectively has to be searched for on a p-dimensional grid. Even if refined search strategies are used (e.g. Wood, 2004) problems are likely to occur when the number of observations is small. At some point a model will have to be evaluated that allows for much flexibility for a large number of the covariates. Then estimation will run into numerical problems.

In the following a new technique for fitting generalized additive models is presented that is designed to work well even if there is a large number of covariates and only (relatively) few observations.

1.2 Generalized additive model boosting

1.2.1 Boosting and gradient descent

The notion of *boosting* was introduced in the Machine Learning community in the context of classification for binary response data. Several weak learners, i.e., models that have prediction performance only slightly above chance level, are combined into a prediction device that is found to have far better performance than any of the single weak learners (Schapire, 1990). AdaBoost (Freund and Schapire, 1996), one of the most prominent boosting algorithms, achieves this by repeatedly fitting a weak learner to the re-weighted data, where points that were misclassified in the last step receive higher weights. The models fitted in each step then are combined into a final classifier by a weighted majority vote.

Friedman et al. (2000) cast boosting into the framework of gradient descent in function space (see also Friedman, 2001; Hastie et al., 2001; Bühlmann and Yu, 2003). The objective there is to estimate a function F that minimizes the expected loss

$$E\left(L(y, F(x))\right)$$

where $y' = (y_1, \ldots, y_n)$, $x' = (x'_1, \ldots, x'_n)$ and L is a loss function. In a stepwise procedure (starting with $\hat{F}^{(0)} = \hat{f}^{(0)}$) in each step $l = 1, \ldots, m$ the negative gradient vector

$$-g_i^{(l)} = \left[\frac{\partial L(y, F(x_i))}{\partial F(x_i)} \right]_{F(x) = \hat{F}^{(l-1)}(x)} \quad i = 1, \ldots, n$$

is calculated and a learner is fitted to it, resulting in the estimate $\hat{f}^{(l)} = (\hat{f}^{(l)}(x_1), \ldots, \hat{f}^{(l)}(x_n))'$ which is used for the update

$$\hat{F}^{(l)}(x) = \hat{F}^{(l-1)}(x) + \hat{w}_m \hat{f}^{(l)}$$

where

$$\hat{w}_l = \arg \min_w \sum_{i=1}^{n} L(y_i, \hat{F}^{(l-1)}(x_i) + w \hat{f}^{(l)}(x_i)).$$

When least-squares error

$$L(x, F) = \frac{1}{2} \sum_{i=1}^{n} (y_i - F(x_i))^2$$

is used for the loss function this gradient-descent procedure is called L2Boost (Bühlmann and Yu, 2003). By calculating the gradient

$$-g_i^{(l)} = y_i - \hat{F}^{(l)}(x_i)$$

it can be seen that it is equivalent (with $\hat{w}_l = 1$) to repeated fitting of the residuals. In the following this idea is adapted to the generalized linear model framework.

1.2.2 Generalized additive models by repeated fitting of residuals

Due to the origin of boosting in the classification context there are several boosting algorithms for binary response data, AdaBoost (Freund and Schapire, 1996) and LogitBoost (Friedman et al., 2000) being two prominent examples. For Gaussian responses L2Boost (Bühlmann and Yu, 2003) can be used, despite also having been developed for classification tasks. LogitBoost and L2Boost both are based on the gradient descent approach reviewed in the previous section. Thus, when

11

models for other exponential family models (e.g. a Poisson model for counting responses) are needed one could try to adapt the gradient approach for these. The downside to this is that it has to be done for each type of response separately. For developing a general boosting procedure therefore in the following not a gradient approach is used but the idea of repeated fitting of the residuals (seen for L2Boost) is extended to generalized additive models.

First the notion of repeated fitting of residuals is generalized to the generalized linear model framework, using any kind of learner $\eta(x, \gamma)$ that provides a mapping $f_\gamma : \mathbb{R}^p \to \mathbb{R}$ of a covariate vector x to a predicted value, specified by the parameter vector γ. This includes learners such as regression trees (Breiman et al., 1984), that incorporate several covariates in every boosting step. Later the focus will be on learners such as penalized B-splines that use only one covariate in each step. It will be shown that this results in fitting a generalized additive model.

The GAMBoost algorithm proposed, using the link function h and log-likelihood $l(\gamma)$ as introduced in Sections 1.1.1 and 1.1.2, is as follows:

GAMBoost: Estimating generalized additive models by repeated fitting of residuals

Step 1 (Initialization): Initialize $\hat{\eta}_{(0)}(x)$ to some starting value, e.g. an initial intercept.

Step 2 (Iteration): For $l = 1, \ldots, m$ fit the model

$$\mu_i = h(\hat{\eta}_{(l-1)}(x_i) + \eta(x_i)) \tag{1.9}$$

to data (y_i, x_i), $i = 1, \ldots, n$ by maximizing the log-likelihood $l(\gamma)$, where $\hat{\eta}_{(l-1)}(x_i)$ is treated as an offset and $\eta(x_i)$ is estimated by the learner $\eta(x_i, \gamma_l)$. Set $\hat{\eta}_{(l)}(x_i) = \hat{\eta}_{(l-1)}(x_i) + \hat{\eta}(x_i, \hat{\gamma}_l)$ and update

$$\hat{F}^{(l)}(x_i) = h(\hat{\eta}_{(l-1)}(x_i) + \hat{\eta}(x_i, \hat{\gamma}_l)).$$

The intercept term used for a GAMBoost model deserves some special consideration for learners that use centered estimates to avoid uniqueness problems. To avoid difficulties that might occur when such a learner is applied to an un-centered target, an intercept covariate $x_{i0} = 1$ with design vector $z'_{i0} = (1)$ is introduced. The parameter estimate for this component then is updated before each boosting step to provide the actual learner with a centered target to be fitted.

The final estimate $\hat{F}^{(m)}(x)$ that results from the GAMBoost algorithm

$$\hat{F}^{(m)}(x) = h\left(\sum_{l=0}^{m} \hat{\eta}_{(l)}(x)\right),$$

in contrast to other boosting procedures, is not based on a (weighted) sum of the fitted responses $h(\hat{\eta}_{(l)}(x))$ but on a sum of the fitted predictors $\hat{\eta}_{(l)}$. For a Gaussian response with the canonical link, i.e., when h is the identity, these two cases coincide. GAMBoost then is identical to L2Boost.

As noted above, besides choosing the type of learner one has to decide what number of covariates is to be included by the learner fitted in each boosting step (1.9), i.e., for what number of covariates the contribution to the final model $\hat{F}^{(m)}(x)$ is updated. For example, when regression trees with only two terminal nodes (called stumps) are used, only the contribution of one covariate is updated in each step, while for trees with more than two terminal nodes the contribution of several covariates is adjusted. So in each step l there is a fixed-size set

$$V_l \subset \{1, \ldots, p\}$$

of indices of covariates which are considered for an update. It is chosen such that the resulting log-likelihood $l(\hat{F}^{(l)}(x))$ is maximal.

When several covariates are included in each boosting step it is possible to allow for interactions between several covariates, but then the final model may no longer be additive in the contribution of each single covariate. This for example is likely to be the case when regression trees learners with more than two terminal nodes are used. As a fit of a generalized additive model is wanted, investigations are restricted to updates of only one covariate per boosting step in the following. This *componentwise* GAMBoost approach automatically results in a generalized additive model, because the contribution of each step $\hat{\eta}_{(l)}(x)$ can be attributed to one specific covariate. The predictor $\hat{\eta}_i = g(\hat{F}^{(m)}(x_i))$ (where g is the inverse of h) can be decomposed into

$$\hat{\eta}_i = \hat{\eta}_{(0)} + \sum_{j=1}^{p} \hat{f}_j(x_{ij}) \tag{1.10}$$

where

$$\hat{f}_j(x_{ij}) = \sum_{l=1}^{m} I(i \in V_l) \cdot \hat{\eta}_{(l)}((0, \ldots, 0, x_{ij}, 0, \ldots, 0)')$$

with $I(expression)$ being the indicator function that is equal to 1 when *expression* is true and 0 otherwise. Comparing (1.10) to the generalized additive model (1.3) it is obvious that componentwise GAMBoost results in fitting a generalized additive model.

One of the immediate advantages of decomposition (1.10) is that the contribution of each covariate j to the final model can easily be illustrated by plotting $\hat{f}_j(x_{ij})$, regardless of the learner used.

For learners such as P-splines that be can written in the form

$$\eta(x_{ij}, \gamma) = z'_{ij}\gamma \qquad (1.11)$$

with a design vector z_{ij} built from x_{ij}, the individual covariate components $\hat{f}_j(x_{ij})$ of the decomposition (1.10) can be expressed as

$$\hat{f}_j(x_{ij}) = z'_{ij}\hat{\gamma}_j$$

where

$$\hat{\gamma}_j = \sum_{l|j \in V_l} \hat{\gamma}_{V_l} \qquad (1.12)$$

and $\hat{\gamma}_{V_l}$ are the parameter estimates from the boosting steps where the contribution of covariate j has been updated. Therefore the estimates resulting from GAMBoost can be used as if they would be from a one-step estimation. Note that when no basis expansion is used, i.e., $z'_{ij} = (x_{ij})$, this is equivalent to regularized estimation of a (generalized) linear regression model (Tutz and Binder, 2005a).

1.2.3 Specific learners

In the following two types of learners for the GAMBoost algorithm, P-splines and penalized stumps, are introduced and compared. As both use only a single covariate in each boosting step notation is simplified. While in the last section "$\eta(x_i, \gamma)$" was used for a learner including the selection of the covariate that results in the largest likelihood increase, in this section $\eta(x_{ij}, \gamma)$ is used for a learner applied to a covariate j that is to be evaluated with respect to likelihood improvement or has already been selected.

P-splines for the estimation of smooth functions have already been reviewed in Section 1.1.2. Using the form (1.11) with the basis expansion $z'_{ij} = (B_1^{(j)}(x_{ij}), \ldots,$

$B_M^{(j)}(x_{ij}))$ the penalized score function and Fisher matrix used for Fisher scoring (1.8) in the context of a GAMBoost step are

$$s_p(\gamma) = Z_j' D(\gamma) \Sigma(\gamma)^{-1}(y - \mu) - \lambda P \gamma$$

and

$$F_p(\gamma) = Z_j' W(\gamma) Z_j + \lambda P$$

where $y' = (y_1, \ldots, y_n)$, $\mu' = (\mu_1, \ldots, \mu_n)$, $Z_j' = (z_{1j}, \ldots, z_{nj})$, $\Sigma(\gamma) = Diag(\sigma_1^2, \ldots, \sigma_n^2)$, $\sigma_i^2 = \mathrm{var}_\gamma(y_i)$, $D(\gamma) = Diag(\partial h(\eta_i)/\partial\eta, \ldots, \partial h(\eta_n)/\partial\eta)$, $W(\gamma) = D(\gamma) \Sigma(\gamma)^{-1} D(\gamma)$, $\mu_i = h(\eta_i)$, $\eta_i = \hat{\eta}_i^{(l-1)} + z_{ij}' \gamma$ and P takes the form of the individual penalty matrices P_j introduced in Section 1.1.2. Only one step of Fisher scoring is used because any further adjustment of the parameter estimate that might be necessary can also be done in later boosting steps (as can be seen from (1.12)). In addition a vector of zeros $\hat{\gamma}_{Vi0}$ is used as a starting value for the estimate of the parameter vector, which seems reasonable. The Fisher scoring step for the estimation of the parameter vector of a P-spline learner for one GAMBoost step then becomes

$$\hat{\gamma}_{Vi} = (Z_j' W(\hat{\gamma}_{j0}) Z_j + \lambda P)^{-1} Z_j' W(\hat{\gamma}_{j0}) D(\hat{\gamma}_{j0})^{-1}(y - \mu). \qquad (1.13)$$

Regression trees with only two terminal nodes are popular learners for boosting applications (see e.g. Friedman et al., 2000). In each boosting step for a covariate to be evaluated a split point δ is chosen that divides the observations into two groups. For each group the (weighted) mean value of the response to be fitted is taken for prediction. Often that prediction is shrunken towards the overall mean to obtain a weak learner. In the following an alternative method is presented, called *penalized stumps*. For each split point δ and each variable j pseudo-variables $x_{ij,\delta}$ are created, with design vector $z_{ij,\delta} = (z_{ij,\delta,1}, z_{ij,\delta,2})$, where $z_{ij,\delta,1} = I(x_{ij} \leq \delta)$, $z_{ij,\delta,2} = I(x_{ij} > \delta)$ and I is the indicator function with $I(expression) = 1$ if *expression* is true and $I(expression) = 0$ if *expression* does not hold. The tree stump learner then takes the form

$$\eta(x_{ij,\delta}, \gamma) = z_{ij,\delta}' \gamma$$

with parameter vector $\gamma' = (\gamma_1, \gamma_2)$, where γ_1 and γ_2 are the predicted values assigned to the two groups of observations. To obtain estimates for the latter, one-step penalized Fisher scoring of the form (1.13) is used. The resulting improvement of log-likelihood then is used to select one of the pseudo-variables in each boosting step, thus choosing a covariate *and* a split point. For shrinking the

15

predicted values for the two groups towards the overall mean a penalty matrix of the form

$$P_{stump} = \begin{pmatrix} 1 & -1 \\ -1 & 1 \end{pmatrix}$$

is used. With $\lambda = 0$ one obtains common tree stumps as demonstrated in the following. By using the specific structure of the design matrix $Z_{j,\delta}$ one may derive a more explicit form of the estimation step (1.13) yielding

$$\hat{\gamma}_{V_i} = \frac{1}{(\bar{F}_{x_{ij}\leq\delta} + \lambda)(\bar{F}_{x_{ij}>\delta} + \lambda) - \lambda^2} \cdot \begin{pmatrix} (\bar{F}_{x_{ij}>\delta} + \lambda)\bar{s}_{x_{ij}\leq\delta} + \lambda\bar{s}_{x_{ij}>\delta} \\ (\bar{F}_{x_{ij}\leq\delta} + \lambda)\bar{s}_{x_{ij}>\delta} + \lambda\bar{s}_{x_{ij}\leq\delta} \end{pmatrix}$$

where

$$\bar{F}_{x_{ij}\leq\delta} = \sum_{x_{ij}\leq\delta} \frac{1}{\hat{\sigma}_i^2}\left(\frac{\partial h(\hat{\eta}_i)}{\partial\eta}\right)^2, \quad \bar{F}_{x_{ij}>\delta} = \sum_{x_{ij}>\delta} \frac{1}{\hat{\sigma}_i^2}\left(\frac{\partial h(\hat{\eta}_i)}{\partial\eta}\right)^2$$

$$\bar{s}_{x_{ij}\leq\delta} = \sum_{x_{ij}\leq\delta} \frac{1}{\hat{\sigma}_i^2}\frac{\partial h(\hat{\eta}_i)}{\partial\eta}(y_i - \hat{\mu}_i), \quad \bar{s}_{x_{ij}>\delta} = \sum_{x_{ij}>\delta} \frac{1}{\hat{\sigma}_i^2}\frac{\partial h(\hat{\eta}_i)}{\partial\eta}(y_i - \hat{\mu}_i).$$

It can be seen that for a Gaussian response with h being the identity and $\lambda = 0$ estimation reduces to calculation of the mean of the residuals $(y_i - \hat{\mu}_i)$ for observations with $x_{ij} \leq \delta$ and observations with $x_{ij} > \delta$. This is equivalent to common tree stump estimation (see Breiman et al., 1984).

Figure 1.2 shows exemplary GAMBoost fits to binary response data with $p = 5$ covariates and $n = 100$ observations generated from model (1.20) (with parameter $c_n = 3$; discussed in detail later). There are three informative covariates, one having linear influence, the other two being non-linear in their influence on the predictor. The upper panels show a GAMBoost fit with P-spline learners with 20 equidistant knots and penalty $\lambda_{splines} = 30$ and the lower panels a fit with penalized tree stumps with penalty $\lambda_{stumps} = 2$. Although both fits (solid lines) track the true covariate influence functions (broken lines) closely (with the P-spline fit being slightly dampened compared to the true function) the fits based on P-splines are visually more satisfying.

The left panel of Figure 1.3 shows how the mean squared error (MSE) of η depends on the number of boosting iterations for P-spline learners (solid line) and penalized stump learners (broken line) for this example. The dash-dotted line shows the mean MSE for unpenalized stump learners. It can be seen that overfitting occurs earlier when no penalty is used and the minimal value obtained during iterations is rather large. Comparing this to the mean MSE for penalized stumps (broken line) it can be seen that penalization can effectively prevent early

16

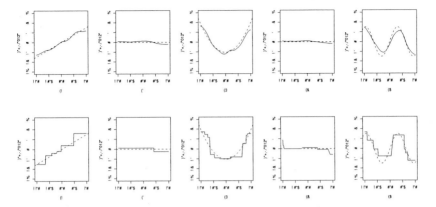

Figure 1.2: GAMBoost fit to simulated data with P-spline (upper panels) and penalized tree stump (lower panels) learners. The broken lines indicate the true covariate influence functions and the solid lines indicate the model fit.

overfitting when tree stump learners are not weak enough. Penalizing stumps also result in a smaller minimum MSE compared to unpenalized stumps. The MSE for both types of tree stumps is consistently larger than for P-spline learners.

The better performance of P-spline learners might be due to the special nature of the simulated data (smooth true functions and no categorical predictors). Therefore binary response data with the same setup as in the previous example was generated but this time using a step function

$$f_{step}(x) = \begin{cases} -1 & \text{for } x \leq 0 \\ 1 & \text{for } x > 0 \end{cases}$$

and a predictor given by

$$\eta_i = c_n(0.5 f_{step}(x_{i1}) + 0.25 f_{step}(x_{i3}) + f_{step}(x_{i5})).$$

The right panel of Figure 1.3 shows the mean MSE for GAMBoost with P-splines and penalized and unpenalized stumps. The penalty parameters used for P-splines and stumps are the same as in the last example. Again unpenalized stumps show early overfitting that can be prevented to some extent by penalization. The minimum MSE achieved by penalized tree stump learners in this case is smaller than that resulting when P-spline learners are used. This does not come as a

17

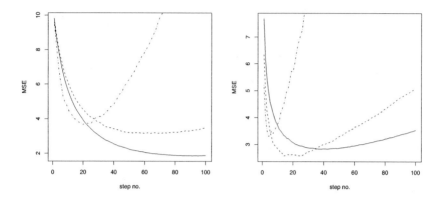

Figure 1.3: Mean curves of mean squared error (MSE) for 20 repetitions of data generation and model fit plotted against the number of boosting steps for data with smooth influence of predictors (left panel) and predictor influence via step functions (right panel) (GAMBoost with penalised B-splines: solid lines; with penalized stumps: broken lines; with unpenalized stumps: dash-dotted lines).

surprise as this second example is designed to favor tree learners compared to learners that use smooth functions such as P-splines.

1.3 Model characteristics

1.3.1 Selection of parameters and model complexity

The GAMBoost algorithm described above has two flexible parameters, the amount of penalization λ and the number of boosting steps m. Both determine the complexity of the resulting model $\hat{F}_{\lambda,m}(x)$. Figure 1.4 shows the mean MSE of GAM-Boost with P-spline learners in the course of the boosting steps for data from the same model that was used for Figure 1.3, but this time for various values of the penalty parameter λ. It can be seen that the penalty parameter can be varied over a wide range and still, similar performance is obtained by using an appropriate number of boosting steps. In addition, the range of the number of boosting steps where good performance can be obtained is rather large. Only when the optimum occurs earlier, it is more difficult to be identified because of

18

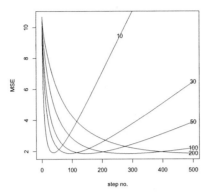

Figure 1.4: Mean curves of mean squared error (MSE) for 20 repetitions of data generation and model fit plotted against the number of boosting steps for data with smooth influence of predictors fitted by GAMBoost with P-spline learners with penalties 10, 30, 50, 100 and 200.

a small range. Therefore in our applications of GAMBoost it is ensured that the optimal number of boosting steps that corresponds to the penalty used is larger than a certain number (say 50) to avoid situations where the learner in a boosting step is not weak enough. Taking this precaution, only one parameter, the number of boosting steps, has to be determined for using GAMBoost. This is in strong contrast to other procedures for fitting generalized additive models, which require selection of a separate smoothing parameter for each covariate. Given an appropriate number of boosting steps GAMBoost if necessary *automatically* assigns different amounts of smoothness to the covariates by using a varying number of updates.

The number of steps to be used for a boosting procedure is chosen such that prediction performance is maximal. The latter typically is determined by cross-validation (see e.g. Friedman et al., 2000; Bühlmann and Yu, 2003). To avoid the computational complexity of this procedure, a criterion for GAMBoost is derived that, based on model complexity, also reflects prediction performance and can be calculated in the course of the boosting steps with less computational complexity compared to cross-validation. For the sake of generality the following notation will reflect the dependence on the penalty γ as well as on the number of boosting steps m.

The criterion that is optimized when the parameters are selected by cross-validation is the expected prediction error of the final GAMBoost estimate $\hat{F}_{\lambda,m}(x)$ for an independent test sample, called *test error* or *generalization error*

$$Err(\lambda, m) = E[L(y, \hat{F}_{\lambda,m}(x))]$$

for some observation (x, y), where the expectation is taken over the joint distribution of x and y and $L(y, \hat{F}_{\lambda,m}(x))$ is some kind of loss function. In the following at first the case of squared error loss $L(y, \hat{F}_{\lambda,m}(x)) = (y - \hat{F}(x))^2$ is examined, which is appropriate for Gaussian response models and then the results are applied to exponential family responses. The generalization error is comprised of an extra-sample error, that arises because the model $\hat{F}_{\lambda,m}(x)$ is evaluated at covariate values not necessarily equal to the x_i it was fitted to, and an in-sample error also called *average predictive squared error* (PSE)

$$PSE(\lambda, m) = \frac{1}{n} \sum_{i=1}^{n} E[y_i^* - \hat{F}_{\lambda,m}(x_i)]^2$$

where y_i^* is a new, independent observation at x_i. A naive approximation to the PSE is the average squared residual

$$ASR(\lambda, m) = \sum_{i=1}^{n} (y_i - \hat{F}_{\lambda,m}(x_i))^2$$

which will be biased because it evaluates performance only for the specific realizations y_i the model has been fitted to, while PSE is based on the whole distribution of y given the x_i. As there may be overfitting to aspects of the specific realization that have no equivalent in the true function $F(x)$ the average squared residual will be an overly optimistic estimate of PSE. When the fitted model $\hat{F}_{\lambda,m}(x)$ can be expressed, using the hat matrix $H_{\lambda,m}$, as $\hat{F}_{\lambda,m}(x) = H_{\lambda,m} y$ with $x' = (x_1, \ldots, x_n)$ and $y' = (y_1, \ldots, y_n)$, and the generating model is $y_i = F(x_i) + \epsilon_i$ with true function $F(x)$, an independent error term ϵ_i with expected value 0 and variance σ^2, then the expected average squared residual is

$$E[ASR(\lambda, m)] = \sigma^2 + \frac{1}{n} \sum_{i=1}^{n} (E[\hat{F}_{\lambda,m}(x_i)] - F(x_i))^2 - \frac{\sigma^2}{n} \cdot tr(2H_{\lambda,m} - H_{\lambda,m} H'_{\lambda,m}).$$

Comparing this to the PSE expressed in terms of the hat matrix

$$PSE(\lambda, m) = \sigma^2 + \frac{1}{n} \sum_{i=1}^{n} (E[\hat{F}_{\lambda,m}(x_i)] - F(x_i))^2 + \frac{\sigma^2}{n} \cdot tr(H_{\lambda,m} H'_{\lambda,m})$$

it can be seen that the expected average squared residual differs from PSE by $2 \cdot tr(H_{\lambda,m})\sigma^2/n$. By adding the latter as a correction term to the average squared residual one obtains *Mallows' C_p* criterion

$$C_p(\lambda, m) = ASR(\lambda, m) + 2 \cdot trace(H_{\lambda,m})\hat{\sigma}^2/n$$

(Mallows, 1973) where for simple linear models the trace of the hat matrix $tr(H_{\lambda,m})$ is equal to the number of covariates included (Efron, 1986). For more details in the context of additive models see Hastie and Tibshirani (1990).

For exponential family responses instead of squared error loss the *deviance*

$$D_i(y_i, \hat{F}_{\lambda,m}(x_i)) = -2\phi(l_i(\hat{F}_{\lambda,m}(x_i)) - l_i(y_i))$$

is used as a loss function. Using deviance on new data or Kullback-Leibler distance the role of PSE is taken by prediction error

$$PE(\lambda, m) = \frac{1}{n} \sum_{i=1}^{n} E[D(y_i^*, \hat{F}_{\lambda,m}(x_i))] \tag{1.14}$$

where again y_i^* is a new, independent observation at x_i. Similar to the relation of average squared residuals and PSE, deviance of the observations

$$D(\lambda, m) = \frac{1}{n} \sum_{i=1}^{n} D_i(y_i, \hat{F}_{\lambda,m}(x_i)) \tag{1.15}$$

will be biased as an estimate of (1.14). Adding $2df\phi/n$, where df is the effective degrees of freedom in the model, makes it asymptotically unbiased (Efron, 1986; Hastie and Tibshirani, 1990). This leads to Akaike's information criterion

$$AIC(\lambda, m) = D(\lambda, m) + 2df\phi/n, \tag{1.16}$$

(Akaike, 1973) which is equivalent to the C_p criterion in the Gaussian case.

For being able to use AIC as a criterion for the selection of the number of GAM-Boost steps the effective degrees of freedom of the final model are needed. Therefore a derivation of the (approximate) hat matrix for GAMBoost is given in the following.

For the learners presented in Section 1.2.3 the predictor $\eta' = (\eta_1, \ldots, \eta_n)$ can be written in the form $\eta = Z\beta$, where the total design matrix $Z = (1_n, Z_1, \ldots, Z_p)$ decomposes into the design matrices for single covariates $Z_j' = (z_{1j}, \ldots, z_{nj}), j =$

$1, \ldots, p$ and an intercept vector where $1'_n = (1, \ldots, 1)$. Similarly the vector β decomposes into $\beta' = (\gamma_0, \gamma'_1, \ldots, \gamma'_p)$ where γ_i, formed by a sum of per-step-parameter vectors (1.12), denotes the parameters for the expansions of the co-variates collected in Z_j. The additive structure is captured in

$$Z\beta = 1 \cdot \gamma_0 + Z_1\gamma_1 + \cdots + Z_p\gamma_p = \eta_0 + f_1 + \cdots + f_p.$$

Let j denote the variable that is selected in the lth iteration, i.e., the candidate set is $V_l = \{j\}$. In the following the notation Z_{V_l} is used instead of Z_j for the design matrix of the selected variable to indicate that the extension to a candidate set with more elements can easily be achieved. The linear predictor of the fitted model has the form

$$\hat{\eta}_{(l)} = \hat{\eta}_{(l-1)} + Z_{V_l}\gamma_{V_l}.$$

With estimate $\hat{\gamma}_{V_{l+1}}$ the total vector after the update is denoted by

$$\hat{\beta}_{(l)} = \hat{\beta}_{(l-1)} + (0, 0', \ldots, \hat{\gamma}'_{V_{l+1}}, \ldots, 0')'$$

where $0' = (0, \ldots, 0)$ is a vector of length M, containing only zeros. The initial value is $\hat{\beta}_{(0)} = (\hat{\gamma}_0, 0, \ldots, 0)$, where $\hat{\gamma}_0 = \hat{\eta}_{(0)}$. The predictor in the lth iteration has the form

$$\hat{\eta}_{(l)} = Z\hat{\beta}_{(l-1)} + Z_{V_l}\hat{\gamma}_{V_l} = Z\hat{\beta}_{(l)}.$$

Estimating γ_{V_l} by one-step Fisher scoring (1.13), with starting value $\hat{\gamma}_{V,0}$ and $W_l = W(\hat{\gamma}_{V,0})$, $\Sigma_l = \Sigma(\hat{\gamma}_{V,0})$, $D_l = D(\hat{\gamma}_{V,0})$ denoting evaluations at value $\hat{\eta} = \hat{\eta}_{(l-1)} + Z_{V_l}\hat{\gamma}_{V,0}$, one obtains

$$\begin{aligned}
\hat{\eta}_{(l)} &= Z\hat{\beta}_{(l-1)} + Z_{V_l}\hat{\gamma}_{V_k} \\
\hat{\eta}_{(l)} - \hat{\eta}_{(l-1)} &= Z_{V_l}\hat{\gamma}_{V_j} \\
&= Z_{V_l}(Z'_j W_l Z_{V_l} + \lambda P)^{-1} Z'_j W_l D_l^{-1}(y - \hat{\mu}_{(l-1)})
\end{aligned}$$

where $\hat{\mu}_{(l-1)} = h(\hat{\eta}_{(l-1)} + Z_{V_l}\gamma_{j0})$.

Taylor approximation of first order $h(\hat{\eta}) = h(\eta) + \frac{\partial h(\eta)}{\partial \eta'}(\hat{\eta} - \eta)$ yields

$$\begin{aligned}
\hat{\mu}_{(l)} &\approx \hat{\mu}_{(l-1)} + D_l(\hat{\eta}_{(l)} - \hat{\eta}_{(l-1)}) \\
&= \hat{\mu}_{(l-1)} + D_l Z_i \hat{\gamma}_{V_l} \\
&= \hat{\mu}_{(l-1)} + D_l Z_{V_l}(Z'_j W_l Z_{V_l} + \lambda P)^{-1} Z'_{V_l} W_l D_l^{-1}(y - \hat{\mu}_{(l-1)}).
\end{aligned}$$

By defining

$$M_l = D_l Z_{V_l}(Z'_j W_l Z_{V_l} + \lambda P)^{-1} Z'_{V_l} W_l D_l^{-1}$$

one obtains, with $\hat{\mu}_{(0)} = M_0 y$,

$$
\begin{aligned}
\hat{\mu}_{(l)} &\approx \hat{\mu}_{(l-1)} + M_l(y - \hat{\mu}_{(l-1)}) \\
&= \hat{\mu}_{(l-1)} + M_l(y - \hat{\mu}_{(l-2)} - (\hat{\mu}_{(l-1)} - \hat{\mu}_{(l-2)})) \\
&\approx \hat{\mu}_{(l-1)} + M_l(y - \hat{\mu}_{(l-2)} - M_{l-1}(y - \hat{\mu}_{(l-2)})) \\
&= \hat{\mu}_{(l-1)} + M_l(I - M_{l-1})(y - \hat{\mu}_{(l-2)}) \\
&\approx \hat{\mu}_{(l-1)} + M_l \prod_{i=0}^{l-1}(I - M_i)y \\
&\approx \hat{\mu}_{(l-2)} + M_{l-1} \prod_{i=0}^{l-2}(I - M_i)\, y + M_l \prod_{i=0}^{l-1}(I - M_i)\, y \\
&\approx \sum_{j=0}^{l} M_j \prod_{i=0}^{j-1}(I - M_i)\, y.
\end{aligned}
$$

So the corresponding approximate hat matrix after step l is given by

$$
H_l = \sum_{j=0}^{l} M_j \prod_{i=0}^{j-1}(I - M_i). \tag{1.17}
$$

When the initial estimate $\hat{\eta}_{(0)}$ is obtained by fitting of an intercept which yields $\hat{\mu}'_0 = (\bar{y}, \ldots, \bar{y})$, M_0 is given by $M_0 = \frac{1}{n} 1_n 1'_n$ where $1'_n = (1, \ldots, 1)$.

The trace of the hat matrix H_l can be used as the effective degrees of freedom. For example the AIC can be calculated after every boosting step to determine online when GAMBoost should be stopped. Note however that when h is not the identity (typically for a non-Gaussian response) the hat matrix is only approximate. Furthermore the selection of a covariate in every boosting step is not reflected. Ignoring the impact of this is equivalent to assuming that a fixed sequence of covariates is used. In Section 1.4.2 a simulation study is used to evaluate if AIC based on the effective degrees of freedom from this hat matrix is working for the selection of the number of GAMBoost steps.

1.3.2 Confidence bands

In the following pointwise confidence intervals for the estimated covariate influence functions are derived. The linear predictor after l iterations decomposes into

$$\hat{\eta}_{(l)} = \hat{\eta}_{(0)} + Z_1\hat{\gamma}_{(l),1} + \cdots + Z_p\hat{\gamma}_{(l),p}$$

where $\hat{\gamma}_{(l),1}$ denotes the cumulated parameter estimates (1.12) for covariate j in step l.

With $j_{l+1} \in \{1,\ldots,p\}$ denoting the variable chosen in the $(l+1)$th step one has

$$Z_j\hat{\gamma}_{(l+1),j} = \begin{cases} Z_j\hat{\gamma}_{(l),j} & j_{l+1} \neq j \\ Z_j\hat{\gamma}_{(l),j} + R_{l+1}(y - \hat{\mu}_{(l)}) & j_{l+1} = j \end{cases}$$

where $R_{l+1} = Z_j(Z_j'W_lZ_j + \lambda P)^{-1}Z_jW_lD_l^{-1}$.

The form reflects that only one component is updated at a time. In closed form one obtains for the jth component

$$Z_j\hat{\beta}_{(l+1),j} = Z_j\hat{\gamma}_{(l),j} + I(j_{l+1} = j)R_{l+1}(y - \hat{\mu}_{(l)})$$

where I is the indicator function with $I(expression) = 1$ if $expression$ is true and $I(expression) = 0$ if $expression$ does not hold. With approximation $\hat{\mu}_{(m)} \approx H_my$ one obtains

$$Z_j\hat{\gamma}_{(l+1),j} \approx Z_{V_l}\hat{\gamma}_{(l),j} + I(j_{l+1} = j)R_{l+1}(I - H_l)y$$

and therefore

$$Z_j\hat{\gamma}_{(l),j} \approx Q_{m,j}y$$

where

$$Q_{l,j} = \sum_{k=0}^{l-1} I(j_{l+1} = j)R_{l+1}(I - H_{(k)}).$$

From

$$c\hat{o}v(Q_{l,j}y) = Q_{m,j}c\hat{o}v(y)Q_{l,j}' \tag{1.18}$$

where $c\hat{o}v(y) = diag(\hat{\sigma}_1^2,\ldots,\hat{\sigma}_n^2)$, approximate pointwise confidence intervals of $\hat{f}_{(l),j} = Z_{V_l}\hat{\beta}_{(l),j} \approx Q_{l,j}y$ are found. For Gaussian response with h being the identity this is equal to the formulation suggested in Hastie and Tibshirani (1990).

1.4 Comparison to alternative fitting methods

1.4.1 GAMBoost and LogitBoost

For binary response data, LogitBoost (Friedman et al., 2000) is a well established general boosting procedure which uses the gradient approach reviewed in Section 1.2.1. In each step the working response $z_i = y_i - p_{(l)}(x_i)/(p_{(l)}(x_i)(1 - p_{(l)}(x_i)))$ is calculated and fitted with weights $w_i = p_{(l)}(x_i)(1 - p_{(l)}(x))$, where $p_{(l)}(x_i)$ is the predicted probability of $y = 1$ based on the past boosting steps. The resulting estimate $f_{(l+1)}$ is used to update the total estimate $\eta_{(l)}(x)$ by

$$\eta_{(l+1)}(x) = \eta_{(l)}(x) + \frac{1}{2}f_{(l+1)}(x)$$

and $p_{(l+1)}$ is obtained by

$$p_{(l+1)}(x) = \frac{\exp(\hat{\eta}_{(l+1)}(x))}{\exp(\hat{\eta}_{(l+1)}(x)) + \exp(-\hat{\eta}_{(l+1)}(x))}.$$

When LogitBoost is used for componentwise boosting, i.e., only one covariate is used in each step, in combination with a learner from Section 1.2.3 the estimation step for covariate j is

$$f_j^{(l+1)} = Z_j(Z_j'\hat{W}_L Z_j + \lambda P)^{-1} Z_j' \hat{W}_L z_{(l)}$$

where $W_l = Diag(p_{(l)}(x_1)(1 - p_{(l)}(x_1)), \ldots, p_{(l)}(x_n)(1 - p_{(l)}(x_n)))$ is a weight matrix and $z'_{(l)} = (z_1, \ldots, z_n)$ is the working response after step l.

The corresponding GAMBoost step for a binary response data using the (canonical) logit link with $\partial h(\eta_i)/\partial \eta = var(y_i)/\phi$ takes the form

$$f_j = Z_j(Z_j'\hat{D}Z_j + \lambda P)^{-1} Z_j'(y - \hat{\mu}).$$

Since for the canonical link $\hat{D} = \hat{W}_L$ and $\hat{W}_L z_{(l)} = (y_1 - p_{(l)}(x_1), \ldots, y_n - p_{(l)}(x_n))'$ it can be seen that with the same starting point (i.e., $\hat{\mu}_i^{(l)} = p_{(l)}(x_i)$) the update is virtually identical for LogitBoost and GAMBoost.

1.4.2 Empirical comparison

Simulated data examples are used to compare the performance of GAMBoost to that of other procedures for function estimation and variable selection in gener-

Table 1.1: Procedures used for fitting generalized additive models to simulated data.

Name	Description	Criterion	Package
base	Generalized linear model (GLM) with intercept term only	-	stats
GLM	GLM using all predictors	-	stats
GLMse	GLM using backward elimination of covariates	AIC	MASS
bfGAM	Generalized additive models (GAM) with back-fitting/local scoring with stepwise selection of model components and degrees of freedom	AIC	gam 0.94
wGAM	GAM with simultaneous estimation of all components incorporating automatic selection of smoothing parameters and backward elimination of covariates	GCV/UBRE	mgcv 1.3-1
wGAMfb	Like wGAM, but allowing for linear terms and using a stepwise selection of covariates starting from an "all linear" model.	GCV/UBRE	mgcv 1.3-1
GB	GAMBoost with P-spline learners with the number of boosting steps chosen by AIC	AIC	own
GB opt	Like GB, but number of steps chosen for optimal prediction performance	prediction	own

alized additive models. The advantage compared to real data is, that the true form of the covariate influence is known and new data can easily be generated to evaluate prediction performance. Only such procedures were used for comparison that explicitly allow for an exponential family response and give estimates for the shape of the contribution of each covariate separately together with confidence intervals.

The simulation study was done in the statistical environment R (R Development Core Team, 2005, Version 2.1.1). Table 1.1 lists the procedures used with a short description, a short name for reference, the R package used and the optimization criterion for selection of smoothing parameters or degrees of freedom. All parameters were set to default values except the maximum number of iterations, which was increased because often the initial number was not sufficient for binary response examples. As indicated in Table 1.1 for generalized linear models the covariates to be included are selected by backward elimination guided by AIC (see e.g. McCullagh and Nelder, 1989). For generalized additive models in addition the amount of flexibility to be used has to be specified for every covariate. For generalized additive models with backfitting/local scoring (bfGAM) (Hastie and Tibshirani, 1990) a stepwise procedure suggested in Chambers and Hastie (1992) is used for selecting the degrees of freedom. Starting from an initial model that includes all covariates as linear predictors, for each predictor an upgrade and a downgrade by one level (with levels "not included", "linear" and "smooth" with 4, 6 and 12 degrees of freedom) is evaluated subsequently, and the predictor is modified such that a maximum decrease of the AIC results. This upgrade/downgrade procedure is repeated until no further decrease of AIC can be achieved. For generalized additive modelling with simultaneous estimation of all components (wGAM and wGAMfb) (Wood, 2004) generalized cross-validation (GCV) for the Gaussian response examples and the unbiased risk estimator (UBRE) for binary and Poisson response examples is used for selection of model components and selection of smoothing parameters (Wood, 2004). While wGAM uses backward elimination, wGAMfb uses an approach similar to that used for bfGAM. Starting from a model that includes all covariates as linear predictors, for each covariate the levels "not included", "linear", and "smooth with degrees of freedom determined by GCV/UBRE" are evaluated in a stepwise procedure until GCV/UBRE can no longer be improved. The two flexible parameters of GAMBoost are the number of boosting steps and the penalty parameter λ. The first is chosen according to the approximate AIC given in (1.16). Based on the reasoning given in Section 1.3.1 the second is chosen by an automatic procedure that starts with a small value for λ and then uses a rather coarse line search to determine a final λ that results in the number boosting steps being chosen to be greater or equal to 50. As weak learners P-splines with 20 equidistant knots and second order penalization are used.

For the simulated data p covariates $x_{i1}, \ldots, x_{ip}$ (with $p \in \{3, 5, 10, 20, 50\}$) are generated, each drawn from a uniform distribution with values between -1 and 1. There are examples with linear influence of covariates with

$$\eta_i = c_n(x_{i1} + 0.7x_{i3} + 1.5x_{i5}) \tag{1.19}$$

and examples with non-linear components

$$\eta_i = c_n(-0.7 + x_{i1} + 2x_{i3}^2 + \sin 5x_{i5}) \tag{1.20}$$

resulting in three informative covariates x_{i1}, x_{i3}, and x_{i5}. For $p < 5$ (1.19) and (1.20) are modified to use informative covariates x_{i1}, x_{i2} and x_{i3}. The distribution the responses y_i are drawn from is a normal distribution $y_i \sim N(\eta_i, 1)$ for the Gaussian response examples, a binomial distribution $y_i \sim B\left(1, \frac{\exp(\eta_i)}{1+\exp(\eta_i)}\right)$ for the binary response examples, and a Poisson distribution $y_i \sim \text{Poisson}(\exp(\eta_i))$ for the Poisson response examples. The parameter c_n which takes three values (0.5, 0.75, and 1 for the Poisson response examples, 1, 2, and 3 for the binary response examples and 0.5, 1, and 3 for the Gaussian response examples) corresponding to "high noise", "some noise", and "low noise" controls the amount of influence of the covariates on the response. With smaller values of c_n it will be harder to identify the shape of the predictor contribution as there is a smaller signal-to-noise ratio.

For each combination of p and c_n 50 training samples of size $n = 100$ were created. For each the performance of the models fitted by the procedures was evaluated on a new test sample of size 1000. As a measure of goodness-of-fit the Kullback-Leibler distance or deviance (1.15) is used as it is equally adequate for the Gaussian, the binary, and the Poisson case. For the GAM procedures the question arises which data sets to take into considerations when warnings concerning convergence are given. Following the suggestions in the wGAM implementation cases were excluded where estimation did not converged and cases where extreme fitted values (close to 0 or 1 for binary data) occurred. If these cases were included the results for the GAM procedures would have been much worse, thus the exclusion was not in favor of GAMBoost.

For application to real data it is often important that no relevant predictor is missed, but nevertheless parsimonious models are selected. To evaluate this aspect for some of the examples the *hit rate* (i.e., the proportion of correctly identified influential variables)

$$\text{hit rate} = \frac{1}{\sum_{j=1}^{p} I(j \in V_{true})} \sum_{j=1}^{p} I(j \in V_{true}) \cdot I(j \in V_{fitted})$$

and the *false alarm rate* (i.e., the proportion of non-influential variables dubbed influential)

$$\text{false alarm rate} = \frac{1}{\sum_{j=1}^{p} I(j \notin V_{true})} \sum_{j=1}^{p} I(j \notin V_{true}) \cdot I(j \in V_{fitted})$$

are reported, where V_{true} is the set of informative predictors ($V_{true} = \{1, 3, 5\}$ in our examples), V_{fitted} is the set of predictors used by the final model and $I(expression)$ is the indicator function that takes the value 1 if *expression* is true and 0 otherwise.

Linear covariate influence

When models such as generalized additive models are used that provide a large amount of flexibility, one has to assure that they are well-behaved, i.e., no overfitting occurs, when the underlying structure of the data is simple. This is the

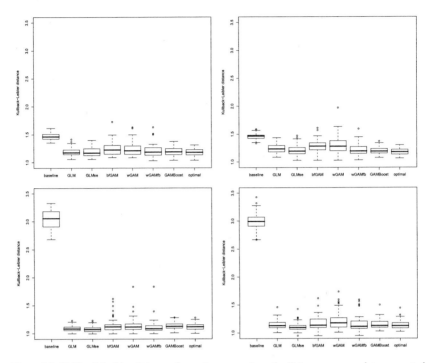

Figure 1.5: Kullback-Leibler distances for various procedures for Poisson response data generated from model (1.19) with linear covariate influence with $c_n = 0.5$ (upper panels) $c_n = 1$ (lower panels) and 5 (left panels) and 10 (right panels) covariates (baseline: GLM with only intercept; GLM: GLM with full model; GLMse: GLM with backward elimination; bfGAM: backfitting GAM with forward/backward; wGAM: Wood GAM with backward elimination; wGAMfb: Wood GAM with forward/backward; GAMBoost: GAMBoost with AIC stopping criterion; optimal: GAMBoost with optimal number of boosting steps). Instances with extreme fitted values were excluded.

motivation for generating data with linear influence of covariates from model (1.19). It will be examined how well superfluous flexibility is controlled by the parameter selection and estimation techniques employed by procedures for fitting generalized additive models.

Figure 1.5 shows the Kullback-Leibler distances obtained from 50 repetitions for Poisson response data for $p = 5$ (left panels) and $p = 10$ (right panels) and two levels of signal-to-noise ratio determined by $c_n = 0.5$ (upper panels) and $c_n = 1$ (lower panels). It can be seen that all procedures improve over the

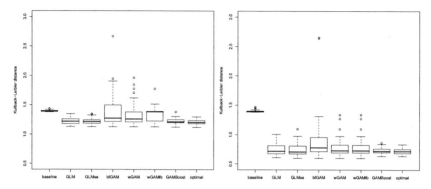

Figure 1.6: Kullback-Leibler distances for various procedures for binary response data generated from model (1.19) with linear covariate influence with $c_n = 1$ (left) and $c_n = 3$ (right) and 5 covariates (baseline: GLM with only intercept; GLM: GLM with full model; GLMse: GLM with backward elimination; bfGAM: backfitting GAM with forward/backward; wGAM: Wood GAM with backward elimination; wGAMfb: Wood GAM with forward/backward; GAMBoost: GAMBoost with AIC stopping criterion; optimal: GAMBoost with optimal number of boosting steps). Instances with extreme fitted values were excluded.

baseline intercept model, the difference being larger for a large signal-to-noise ratio (lower panels). The GLM procedures (GLM and GLMse) perform very well in all examples, as would be expected. For $p = 5$ all GAM procedures and GAMBoost are close contenders, indicating the their superfluous flexibility is controlled adequately. For a larger number predictors and small signal-to-noise ratio the use of bfGAM and wGAM seems to be more problematic, and only GAMBoost and wGAMfb are still well comparable to the performance of GLM. If in addition variability is considered there occurs a slight advantage for GAMBoost. Figure 1.6 shows the results for binary response examples with $p = 5$. The slight advantage of GLM over GAMBoost found for the Poisson examples can no longer be found here. For a large signal-to-noise ratio (right panel) GAMBoost performs among the best, even though there are only five predictors. For a small signal-to-noise ratio (left panel) it clearly outperforms all other GAM procedures, some of which show performance worse than the baseline in many instances. In summary, GAMBoost seems to work well, even if there is only linear covariate structure in the data. The best performance can be seen with many predictors and small signal-to-noise ratio.

Non-linear covariate influence

In the following the behavior of the procedures in situations is explored where there is non-linear covariate influence, i.e., the flexibility provided by generalized additive models is really needed.

Table 1.2 shows the mean Kullback-Leibler distances for Poisson response data generated from model (1.20). The number of data sets that where excluded (non-convergence or extreme fitted values) is given in parentheses, but such problems seem to be very rare for the Poisson examples. GLM procedures appear to perform much worse compared to GAMBoost and all GAM procedures. This indicates that the additional flexibility of the latter is assigned appropriately by the automatic model selection criteria. The GAM procedures perform well for a moderate number of covariates ($p \leq 10$) and even outperform GAMBoost when the signal-to-noise ratio is large. Fits for the GAM procedures no longer can be obtained for a larger number of covariates or the resulting performance is unacceptable. For a small signal-to-noise ratio (with $c_n = 0.5$) GAMBoost dominates regardless of the number of covariates. The difference is relatively small between the results that are obtained for GAMBoost if the number of boosting steps is selected by approximate AIC, and the results obtained if the optimal number of steps is used. This indicates that selection by approximate AIC works rather well, i.e., the approximate hat matrix and the effective degrees of freedom introduced in Section 1.3.1 are useful.

Figure 1.7 shows the hit rates and false alarm rates for GLMse, wGAMfb and GAMBoost that correspond to the results in Table 1.2. The three levels of signal-to-noise ratios are indicated by different line styles. The arrowheads point in the direction of an increasing number of covariates. The results for GLMse (left panel) indicate that with increasing signal-to-noise ratio more predictors are included in the model and thus the hit rate as well as the false alarm rate increases. For wGAMfb (middle panel) the hit rate also increases with increasing signal-to-noise ratio, but the false alarm rate stays approximately the same, being very low for all examples. With an increasing number of predictors the hit rate drops drastically. In contrast use of GAMBoost (right panel) results in a relatively large false alarm rate combined with superior hit rate. With an increasing number of predictors the false alarm rate drops, while the hit rate is not affected as strongly as the one of wGAMfb. The plot for GAMBoost might suggest that an increasing signal-to-noise ratio affects GAMBoost similar to GLMse in that the hit rate and the false alarm rate both increase. The difference is that for GAMBoost there occurs a ceiling effect, because the hit rate cannot increase beyond 1.

Table 1.2: Mean Kullback-Leibler distance for Poisson response data generated from model (1.20) for varying numbers of covariates p and varying amount of predictor influence c_n (base: GLM with only intercept; GLM: GLM with full model; GLMse: GLM with backward elimination; bfGAM: backfitting GAM with forward/backward; wGAM: Wood GAM with backward elimination; wGAMfb: Wood GAM with forward/backward; GB: GAMBoost; GB opt: GAMBoost with optimal number of boosting steps). The number of instances that were excluded because of extreme fitted values or where estimation did not converge is given in parentheses. The best result obtainable for each example is printed in boldface.

c_n	p	base	GLM	GLMse	bfGAM	wGAM	wGAMfb	GB	GB opt
0.5	3	1.484	1.423	1.423	1.283	1.267	1.316	**1.243**	1.219
	5	1.484	1.449	1.442	1.311	1.308	1.346	**1.269**	1.248
	10	1.479	1.525	1.490	1.398	1.472	1.452	**1.310**	1.285
	20	1.471	1.695	1.587	-	-	1.561	**1.346**	1.324
	50	1.449	4.332	2.538	-	-	2.457	**1.429**	1.367
0.75	3	2.027	1.816	1.814	1.229	1.220	1.231	**1.215**	1.193
	5	2.027	1.849	1.840	**1.250**	1.262	1.270	1.252	1.229
	10	2.013	1.968	1.928	**1.328**	1.438 (1)	1.427	1.332	1.308
	20	2.024	2.272	2.188	-	-	1.543	**1.401**	1.386
	50	1.971	6.589	3.861	-	-	3.555	**1.539**	1.506
1	3	3.110	2.639	2.641	**1.158**	1.159	1.158	1.184	1.160
	5	3.110	2.703	2.694	**1.185**	1.201	1.194	1.246	1.217
	10	3.100	2.907	2.874	**1.301**	1.393 (1)	1.332	1.347	1.325
	20	3.117	3.379	3.283	-	-	1.540	**1.479**	1.446
	50	3.079	12.297	8.515	-	-	5.054	**1.755**	1.721

Table 1.3 gives the mean Kullback-Leibler distances for the binary response examples generated from model (1.20). The number of data sets that where excluded (non-convergence or extreme fitted values) is given in parentheses. So the mean Kullback-Leibler distance given in these cells is based on a smaller number of repetitions (having possibly less complicated data structure). An empty cell indicates that no fit at all (due to numerical problems) or not a single acceptable fit (according to the exclusion criterion) could be obtained for a procedure. The most severe problems in this area are seen for wGAM. For $p \geq 10$ in each example for at least half of the 50 repetitions no fit can be obtained. Therefore the results for this procedure are very difficult to compare to the others. Given that there is only one covariate with linear influence in the generating model (1.20), the GLM procedures perform rather well. For a small signal-to-noise ratio they even outperform some of the GAM procedures. This indicates that the latter perform rather poorly for such binary response examples with a low level of information. For $p > 5$ the performance is worse than the fitting of an intercept model, for $p = 5$ it is within the range of the intercept model. This is strongly contrasted by the GAMBoost procedure which (with one exception) strictly performs better than the simple intercept model and is very robust against the inclusion of noise

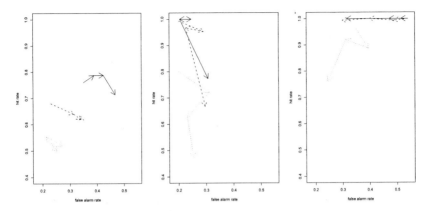

Figure 1.7: Hit rates/false alarm rates for Poisson response data generated from model (1.20) for varying a number of covariates p (connected by arrows) and varying amount of covariate influence ($c_n = 1$: solid lines; $c_n = 0.75$: dash-dotted lines; $c_n = 0.5$: dotted lines) for GLMse (left), wGAMfb (middle) and GAMBoost (right).

variables. The performance is rather the same up to 20 variables, an increase in Kullback-Leibler distance can be seen only for 50 variables where all of the other GAM procedures perform rather poorly. So GAMBoost seems to be very resistant to overfitting. Comparing the performance of GAMBoost with the number of steps selected by approximate AIC to the results that can be obtained with the optimal number of boosting steps it can be seen that, similar to the Poisson response examples, the AIC is working well as a criterion for the selection of the number of boosting steps.

Comparing the results for the Poisson response examples to that for the binary response examples, it becomes clear that GAMBoost, while working well for the former, performs especially well with the latter. As the advantage of using GAMBoost is larger when the signal-to-noise ratio is low one might conjecture that its performance benefit for binary response examples is due to the relatively low amount of information in such data. While a Poisson response contains more information compared to a binary response, the maximal amount of information from the predictor η_i is available for a Gaussian response. Therefore some Gaussian response examples are examined. Note that in contrast to the more general exponential family case there is large number of model building techniques for Gaussian responses. Therefore no attempt for a comprehensive discussion is given, but just a comparison to the Poisson examples. In addition, as the use

Table 1.3: Mean Kullback-Leibler distance for binary response data generated from model (1.20) for varying numbers of covariates p and varying amount of covariate influence c_n (base: GLM with only intercept; GLM: GLM with full model; GLMse: GLM with backward elimination; bfGAM: backfitting GAM with forward/backward; wGAM: Wood GAM with backward elimination; wGAMfb: Wood GAM with forward/backward; GB: GAMBoost; GB opt: GAMBoost with optimal number of boosting steps). The number of instances that were excluded because of extreme fitted values or where estimation did not converge is given in parentheses. The best result obtainable for each example is printed in boldface.

c_n	p	base	GLM	GLMse	bfGAM	wGAM	wGAMfb	GB	GB opt
1	3	1.400	1.371	1.370	1.364	1.351	1.400	**1.293**	1.258
	5	1.400	1.396	1.386	1.538	1.453 (1)	1.407	**1.319**	1.284
	10	1.398	1.492	1.452	2.197	1.507 (30)	1.414 (4)	**1.356**	1.319
	20	1.399	1.707	1.538	-	-	1.506 (24)	**1.380**	1.333
	50	**1.395**	4.554 (38)	2.564 (35)	-	-	4.264 (46)	1.489	1.365
2	3	1.400	1.305	1.306	1.141 (1)	1.097 (4)	1.304 (1)	**0.984**	0.965
	5	1.400	1.333	1.325	1.472 (3)	1.341 (14)	1.362 (7)	**1.028**	1
	10	1.400	1.424	1.378	1.711 (12)	1.509 (45)	1.358 (28)	**1.076**	1.057
	20	1.397	1.646	1.496	-	-	1.398 (49)	**1.116**	1.099
	50	1.395	4.836 (43)	2.617 (40)	-	-	2.547 (48)	**1.244**	1.193
3	3	1.400	1.276	1.274	0.961 (7)	0.892 (14)	1.045 (12)	**0.796**	0.767
	5	1.400	1.298	1.286	1.044 (13)	**0.826** (35)	1.048 (29)	0.832	0.801
	10	1.396	1.372	1.333	1.574 (27)	0.910 (49)	1.211 (45)	**0.862**	0.844
	20	1.397	1.641	1.471	-	-	-	**0.920**	0.903
	50	1.397	3.648 (48)	2.762 (42)	-	-	-	**1.013**	0.991

of AIC is problematic for a Gaussian response when error variance is unknown, procedures that depend on it are not applied. Table 1.4 gives the mean Kullback-Leibler distances for the remaining procedures (wGAM also not being included) for data generated from model (1.20). It can be seen that wGAMfb works well for examples with a small to medium number of predictors and a medium to large signal-to-noise ratio, and even outperforms GAMBoost with the optimal number of boosting steps. For a small signal-to-noise ratio or a large number of predictors GAMBoost outperforms wGAMfb (distinctly in several examples). This enforces the notion that GAMBoost is especially well suited for examples with a small signal-to-noise ratio and a large number of predictors.

Impact of variable selection vs. smoothing parameter selection

For the GAM procedures in contrast to GAMBoost, the selection of model components is performed separately from the selection of the amount of smoothness to be used for each component. Any performance differences to GAMBoost therefore can result from both sources. In addition the results shown in Figure 1.7

Table 1.4: Mean Kullback-Leibler distance for Gaussian response data generated from model (1.20) for varying numbers of covariates p and varying amount of covariate influence c_n (base: GLM with only intercept; GLM: GLM with full model; GLMse: GLM with backward elimination; wGAMfb: Wood GAM with forward/backward; GB opt: GAMBoost with optimal number of boosting steps). The best result for each example is printed in boldface.

c_n	p	base	GLM	GLMse	wGAMfb	GB opt
0.5	5	1.313	1.273	1.262	1.161	**1.123**
	10	1.312	1.351	1.311	1.286	**1.153**
	20	1.321	1.518	1.424	1.533	**1.201**
1	5	2.248	1.959	1.939	**1.155**	1.161
	10	2.242	2.079	2.014	1.253	**1.209**
	20	2.249	2.345	2.190	1.497	**1.294**
3	5	12.197	9.259	9.202	**1.203**	1.274
	10	12.123	9.825	9.506	**1.309**	1.368
	20	12.159	11.107	10.255	1.510	**1.495**

indicate that there is a difference in the trade-off between hit rate and false alarm rate. In the following it is investigated to what extent performance differences result from such variable selection characteristics, and to what extent they result from (in-)adequate amounts of smoothness or bad fits for the model components. Exemplarily the wGAM procedure is taken and backward elimination of variables is replaced by GAMBoost variable selection, i.e., GAMBoost with the optimal number of boosting steps is used to determine the covariates to be included and then wGAM is fitted to these.

Table 1.5 shows the difference of the mean Kullback-Leiber distances of this procedure to wGAMfb and GAMBoost for instances where fits for all three could be obtained. The table gives only examples where in more than half of the instances for all three procedure the model fit succeeded. The number of instances that were excluded from the remaining examples is given in parentheses. Positive values for the Kullback-Leibler mean difference indicate better, negative values worse performance.

For the Poisson response examples the performance of wGAM with GAMBoost variable selection is closer to GAMBoost for a small signal-to-noise ratio and a large number of variables. As these are also the examples where wGAMfb performs worse than GAMBoost one might conjecture that this performance difference between the latter two procedures in these example is mainly due to variable selection. Nevertheless there is still some difference between the performance of wGAM with GAMBoost variable selection and the performance of GAMBoost. This indicates that wGAM fails to some extent in appropriately fitting the selected components. For the binary response examples this problem seems to be

Table 1.5: Performance differences of wGAM with variable selection by GAMBoost compared to wGAMfb and GAMBoost for Poisson and binary response examples for a varying number of covariates p and a varying amount of covariate influence c_n. The number of instances that were excluded because of extreme fitted values or where estimation did not converge for one of the three procedures is given in parentheses. Positive values indicate better performance.

c_n	p		diff wGAMfb	diff GAMBoost
Poisson response examples				
0.5	3	(0)	0.055	-0.018
	5	(0)	0.056	-0.021
	10	(0)	0.088	-0.053
	20	(5)	0.156	-0.069
	50	(10)	0.964	-0.007
0.75	3	(0)	0.011	-0.005
	5	(0)	0.005	-0.012
	10	(1)	0.062	-0.031
	20	(17)	0.115	-0.073
1	3	(0)	-0.001	0.025
	5	(0)	-0.010	0.042
	10	(1)	-0.018	-0.002
Binary response examples				
1	3	(0)	0.058	-0.049
	5	(0)	-0.043	-0.131
	10	(4)	-0.077	-0.135
2	3	(4)	0.199	-0.116
	5	(13)	0.039	-0.287
3	3	(14)	0.134	-0.109

more severe, because wGAM with GAMBoost variable selection never comes close to GAMBoost performance.

1.5 Real data

For analyzing patterns of admission/discharge of patients at a hospital one important parameter is the number of readmissions after an initial stay at the hospital. Variables are needed that are connected to the number of future admissions to optimize initial treatment and the planning after discharge. In order to identify and explore such relations, techniques are wanted that provide a graphical representation and do not impose too many restrictions. Additionally the focus lies on homogeneous and therefore often small groups of patients, which causes problems with many complex procedures. In this situation GAMBoost with Poisson response offers flexible modelling for small datasets and helps with the selection

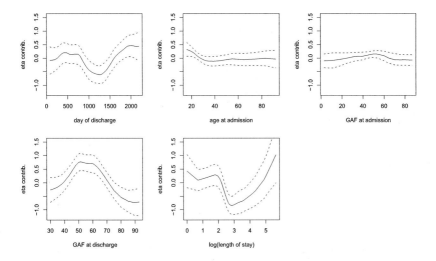

Figure 1.8: Estimated effects of several patient variables collected at the first stay at a psychiatric hospital on the number of readmissions within three years after discharge (estimates: solid lines; pointwise confidence bands: broken lines).

of useful variables. Its ability to cope with a small signal-to-noise ratio is an additional benefit in this instance as the variables available for predicting the number of readmissions may not contain much information. To illustrate the usefulness of GAMBoost it is applied to data from a German psychiatric hospital that in parts has already been shown in Figure 1.1. There are 94 patients with substance abuse that have been treated in the specific hospital within a time range of seven years. The response variable to be investigated is the number of readmissions within a time of three years after discharge. The metric variables available that are likely to carry information with respect to readmissions are "age at admission", (logarithm of) "length of stay (days)" and the "level of functioning" of the patient (GAF for "Global Assessment of Functioning") at admission and at discharge (with larger values indicating a better state of health). Because the patients' data have been collected over a longer period the date of discharge is also included. Further the binary variables "gender" (1:female, 0:male), "compulsory hospitalization" (1:yes, 0:no) and "comorbidity" (1:yes, 0:no) are given. They enter into the per-step selection as predictors with one basis function and zero penalty.

Figure 1.8 shows the GAMBoost estimates for number of readmissions being modelled as a Poisson response with P-spline learners and the number of boosting steps selected by AIC. There seems to be an increased tendency to readmission for younger people (younger than 30). The estimate for "length of stay" shows a drop-off in the medium range with high readmission rates for short-time patients and a tendency to increase for long-term treatment. Looking at the "level of functioning" variables it comes as a surprise that the GAF at admission does not seem to be of high relevance with respect to the number of readmissions. Nevertheless there seems to be a reduction of the number of readmissions with increasing GAF score (at discharge) if it is above 50. This is a plausible result as patients with better health at discharge are less likely to need treatment soon again. The estimate for the variable "compulsory hospitalization" is -0.531 (standard deviation: 0.248), indicating that this type of patient is distinct with respect to the number of readmissions. For the other categorical variables there does not seem to be a significant effect ("sex": estimate 0.108, standard deviation 0.114; "comorbidity": not selected in any boosting iteration).

1.6 Discussion

The newly introduced GAMBoost technique for fitting generalized additive models has been shown to work well for a broad class of examples. It competes well with traditional procedures for a small number of covariates, but its strengths lie in applications with many covariates. The search for optimal smoothing parameters in a p-dimensional space is reduced to the selection of the number of boosting steps and the penalty parameter γ, where the latter might be chosen rather coarsely. The amount of flexibility that is provided for each covariate is automatically adapted by selection of the variables to be updated in each boosting step. Compared to other boosting procedures such as LogitBoost the approximate effective degrees of freedom allow for selection of the number of boosting steps using less computational resources than that required for cross-validation for example. In addition to data with many covariates GAMBoost also performs well when the underlying structure is simple, for in such cases its flexibility is regulated appropriately, or when there is a low level of information (e.g. with a small signal-to-noise ratio or binary response data). It therefore seems to be well suited as a primary exploratory tool for data with unknown structure. The only critical assumption is that of additivity in the covariate influence, but extension to include interactions should be straightforward. Another kind of extension allowing for more flexible models would be to adapt GAMBoost to varying-coefficient models (Hastie and Tibshirani, 1993).

Chapter 2

Flexible modelling of discrete failure time

In the previous chapter a new estimation technique for generalized additive models was introduced, allowing for flexible models with non-linear additive influence of covariates on an exponential family response. This chapter will focus less on new estimation techniques but rather on a new model class. Flexible survival time models that allow for non-parametric estimation of covariate effects based on the generalized linear model framework will be discussed. Besides dealing with other issues that are specific to survival data, e.g. censoring, time as a covariate is given a special role. In the style of varying-coefficient models (Hastie and Tibshirani, 1993) time-varying effects of covariates are considered. Variation over time is not only allowed for covariates with linear influence, but also for non-parametrically estimated effects. Such an approach brings high flexibility to the modelling of survival time data, although it also leads to the danger of model misspecification and overfitting. Therefore a model selection criterion based on effective degrees of freedom is given. Simulated data sets are used explore to what extent such a criterion is useful to identify adequate models.

In Section 2.1.1 the classical framework for survival time models is reviewed with an emphasis on discrete time survival data. In Section 2.1.2 the new class of models is introduced. Section 2.2 investigates its performance with simulated data and Section 2.3 illustrates its application to real data. Section 2.4 gives a final discussion and sketches further possibilities for development.

2.1 Theory

2.1.1 Framework for survival models

The basis for survival time modelling is formed by observations of the form $(t_i, x_{it}), i = 1, \ldots, n$, where t_i indicates the survival time or maximum observed time for a subject i under investigation and the covariate vector $x'_{it} = (x_{it1}, \ldots, x_{itj})$ contains the values of p possibly time-dependent covariates. In classical models, for example the proportional hazards model by Cox (1972), survival time is taken to be a continuous random variate. Kalbfleisch and Prentice (2002) and Lawless (2002) give an overview of the enormous amount of research that has been directed at such models for continuous time. A continuous time varying-coefficient model with time as a covariate that smoothly regulates the effect of other covariates has been introduced by Hastie and Tibshirani (1993). They use a modified partial likelihood algorithm for estimation. In the following the focus will instead be on discrete time survival models.

Often there is only a limited number of follow-ups on the subjects under investigation and therefore each observation occurs only in a certain time interval. Let time be divided into k intervals $[a_0, a_1), [a_1, a_2), \ldots, [a_{q-1}, a_q), [a_q, \infty)$ where $q = k-1$. Time T may also be measured directly on a discrete scale, e.g. in days, weeks or months. Therefore, the discrete time random variable $T \in \{1, \ldots, k\}$ is observed, with $T = t$ indicating failure at time t or in interval $[a_{t-1}, a_t)$.

The focus lies on modelling the hazard function

$$\lambda(t|x_{it}) = P(T = t|T \geq t, x_{it}), \quad t = 1, \ldots, q, \tag{2.1}$$

which completely determines the distribution of T given the covariates in x_{it}. Alternatively, one could directly investigate regression models for T, but often for some of the observations the value T_i is not known, only that it exceeds a certain observation interval indicated by $C_i \in \{1, \ldots, k\}$. This phenomenon is called censoring and the observed time t_i is given by $t_i = \min(T_i, C_i)$. Thus an indicator δ_i with

$$\delta_i = \begin{cases} 1 & \text{if } T_i < C_i \\ 0 & \text{if } T_i \geq C_i \end{cases} \tag{2.2}$$

is introduced. The record for each subject i ($i = 1 \ldots n$) then is given by

$$(t_i, \delta_i, x_{it}).$$

40

One way to deal with such data is the discrete time logistic model

$$\lambda(t|x_{it}) = h(\eta_{it}) = \frac{\exp(\eta_{it})}{1 + \exp(\eta_{it})} \qquad (2.3)$$

considered for example by Thompson (1977), Arjas and Haara (1987), and Efron (1988). Thompson (1977) demonstrated that for short observation intervals it is similar to Cox's continuous time proportional hazards model. Hamerle and Tutz (1989) give a detailed survey of models for discrete survival data, while Fahrmeir and Tutz (2001) treat them in the framework of generalized linear models. The latter source also describes estimation of flexible models by Kalman filtering (see also Fahrmeir, 1994; Fahrmeir and Wagenpfeil, 1996) and fully Bayesian approaches. Klinger et al. (2000) explore the use of regression trees for the identification of influential covariates in logistic survival models. They also give an algorithm for varying-coefficient models with time based on backfitting. While using a different procedure for estimation, in the following this will be extended by allowing for time-varying *smooth* effects of covariates for discrete time logistic survival models.

2.1.2 Flexible models for discrete time

Model

The term η_{it} in discrete time logistic survival models allows for different levels of flexibility. The *baseline* model

$$\eta_{it} = \beta_{0t} \qquad (2.4)$$

with parameters $\beta_{0t}, t = 1, \ldots, q$ does not allow for any covariate influence, but provides an individual hazard rate parameter β_{0t} for every discrete observation time t. This is equivalent to the unspecified baseline hazard of Cox's proportional hazard model. The flexibility of the baseline hazard can be restricted by using the same hazard rate parameter $\beta_{01} = \ldots = \beta_{0T}$ for every observation time, or by assuming a specific parametric form. The latter approach has often been taken to reduce complexity of models (see e.g. Efron, 1988, for the cubic-linear spline model). Later an estimation technique will be introduced and used that does not require such restrictive assumptions. Given the covariate vector $x_{it}, t = 1, \ldots, q$, the model (2.4) can be extended to incorporate covariate effects resulting in the *parametric model*

$$\eta_{it} = \beta_{0t} + x'_{it}\beta \qquad (2.5)$$

with parameter vector $\beta' = (\beta_1, \ldots, \beta_p)$. This is the model used for example by Thompson (1977). Allowing for time-dependent parameter vector $\beta'_t = (\beta_{t1}, \ldots, \beta_{tp}), t = 1, \ldots, q$ the *time-dependent parametric model*

$$\eta_{it} = \beta_{0t} + x'_{it}\beta_t \tag{2.6}$$

is obtained. Written in the form

$$\eta_{it} = \sum_{j=0}^{p} \beta_{(j)}(t)x_{itj} \tag{2.7}$$

it can be seen to be a varying-coefficients model (Hastie and Tibshirani, 1993) with (unspecified) functions $\beta_{(0)}, \ldots, \beta_{(p)}$ and intercept covariate $x_{it0} = 1$.

The models (2.5) and (2.6) are very restrictive in that they require a linear influence of covariates. In the same way as model (2.7) allows for influence of time by unspecified functions it is desirable to allow for influence of some covariates $z_{itj}, j = 1, \ldots, m$ by unknown smooth functions f_j. In Section 1.1 of the previous chapter generalized additive models (Hastie and Tibshirani, 1986, 1990) have been reviewed, which allow for covariate influence on exponential family responses to be estimated with scatterplot smoothers based on the generalized linear model framework. As discrete time logistic survival models also are embedded in this framework much of the former discussion applies as well. The *smooth model* has the form

$$\eta_{it} = \beta_{0t} + \sum_{j=1}^{m} f_j(z_{itj}) \tag{2.8}$$

with unknown smooth functions $f_j, j = 1, \ldots, p$. Note again that for reasons of clarity the covariates that are modelled with smooth functions are denoted by $z_{itj}, j = 1, \ldots, m$ in contrast to the covariates used for parametric components $x_{itj}, j = 1, \ldots, p$.

In the following the smooth model (2.8) will be extended to include time-dependent smooth effects. A general form would allow for smooth functions $f_j(t, z_{itj})$ of a covariate and time. Instead, the focus in the following will be on a more restricted class of models where the smooth influence of the covariates is moderated by an attenuation parameter. This is useful for the modelling of covariates that are measured only once at time $t = 1$ (or earlier), because the influence of such covariates will often diminish (i.e., it will be attenuated) as time progresses. The *time-dependent smooth model* has the form

$$\eta_{it} = \beta_{0t} + \sum_{j=1}^{m} \gamma_{tj} f_j(z_{itj}) \tag{2.9}$$

where $\gamma_{tj}, t = 1, \ldots, q, j = 1, \ldots, m$ are the attenuation parameters. The only restriction put on the attenuation parameters is that the parameters are assumed to vary as a smooth function of time, and that their value at $t = 1$ is fixed at $\gamma_{1j} = 1$, reflecting that the corresponding covariates have full, un-attenuated influence in the beginning.

In principle one could provide the flexibility of time-varying smooth effects for all covariates and use model (2.9) as a general model for discrete time survival data. Nevertheless, often more parsimonious models are wanted, where several of the covariates have a simple parametric effect. The *general model*

$$\eta_{it} = \beta_{0t} + x'_{it}\beta_t + \sum_{j=1}^{m} \gamma_{tj} f_j(z_{itj}). \tag{2.10}$$

allows for parametric as well as smooth covariate effects that may be fixed or may be changing over time. For example it can be used as a basis for a stepwise model selection procedure that identifies which of the covariates require time-varying smooth modelling. In the following an estimation procedure, pointwise confidence bands and a model selection criterion that allow for such an approach will be devised.

Estimation

Estimation of the parameters and unknown functions in the general model (2.10) is closely related to the estimation of generalized additive models reviewed in Section 1.1.2 of the previous chapter. Ignoring the attenuation parameters γ_{tj} estimation is identical to that of a generalized linear model for the parametric components, and to estimation of a generalized additive model for the smooth components. For the parametric components variation over time can be estimated as a varying-coefficient model (see e.g. Klinger et al., 2000). A combined estimation, including time-varying smooth effects, could be based on backfitting (Hastie and Tibshirani, 1990), but as indicated in the previous chapter simultaneous estimation of all covariate effects is preferred. This is accomplished by using B-spline basis expansions and penalized estimation (Eilers and Marx, 1996; Marx and Eilers, 1998; Eilers and Marx, 2002). While for estimation of generalized additive models basis expansion is used only for covariates with smooth effects, for parametric components with a time-varying effect the time t is also treated as a covariate with corresponding basis expansion. The general model (2.10) then can be written in the form

$$\eta_{it} = \sum_{j=0}^{S} \gamma_{tj} \alpha'_j w_{itj} \tag{2.11}$$

where $w'_{itj} = (w_{it1}, \ldots, w_{itM})$ is a design vector and $\alpha'_j = (\alpha_{j1}, \ldots, \alpha_{jM})$ and $\gamma_{tj}, t = 1, \ldots, q, j = 0, \ldots, S(= p + m)$ are unknown parameters. M denotes the number of basis functions to be used for each component. This could easily be extended to allow for a varying number of basis functions for each component.

For covariates with parametric effect and the intercept, i.e., for $j = 0, \ldots, p$, w'_{itj} contains the covariate value (with $x_{it0} = 1$ for the intercept) multiplied by a basis expansion of t

$$w'_{itj} = x_{itj}(B_1^{(j)}(t), \ldots, B_M^{(j)}(t)),$$

while γ_{tj} is fixed at constant value 1, because attenuation over time is already modelled by the basis expansion. For covariates with smooth effects, i.e., $j = p + 1, \ldots, S$, w_{itj} contains the basis expansion of the covariates

$$w'_{itj} = (B_1^{(f_{(j-p)})}(z_{itj}), \ldots, B_M^{(f_{(j-p)})}(z_{itj})).$$

Note that for parametric components without variation over time no basis expansion is needed, i.e., $w'_{itj} = (x_{itj})$, and that for smooth components without effect of time $\gamma_{itj} = 1$ is used. In the following these special cases will not be treated separately, i.e., it is proceeded as if parameters for variation over time would have to be estimated also for these components.

Parameter estimation is done by maximizing the likelihood given the data. Assuming random censoring the likelihood contribution of a single observation i is

$$L_i = c_i \lambda(t_i|x_{it_i})^{\delta_i} \prod_{t=1}^{t_i-1} (1 - \lambda(t|x_{it})) \tag{2.12}$$

where the constant $c_i = P(C_i > t_i)^{\delta_i} P(C_i = t_i)^{1-\delta_i}$ is considered non-informative and in the following will be dropped. In addition it is implicitly assumed that censoring takes place at the beginning of interval $[a_{t_i-1}, a_{t_i})$. Defining a pseudo-response y_{it} with

$$y_{it} = \left\{ \begin{array}{ll} 1 & \text{subject fails in } [a_{t-1}, a_t) \\ 0 & \text{subject survives in } [a_{t-1}, a_t) \end{array} \right.$$

based on (2.12) the log-likelihood of all observations can be shown to be

$$l = \sum_{t=1}^{q} \sum_{i \in R_t} y_{it} \log \lambda(t|x_{it}) + (1 - y_{it}) \log(1 - \lambda(t|x_{it}))$$

44

where the index set $R_t = \{i | t \leq t_i - (1 - \delta_i)\}$ represents the risk set, i.e., the individuals that are still under risk in interval $[a_{t-1}, a_t)$ (for a derivation see e.g. Fahrmeir and Tutz, 2001). For enforcing smoothness of the estimated functions a penalized likelihood similar to (1.7), employed in the previous chapter, is used. For discrete logistic survival models with time-dependent smooth covariate effects two penalty components are needed, one for the smoothness of the basis expansion estimates, and one for the attenuation parameters γ_{it}. In matrix notation the proposed penalized log-likelihood is

$$l_p = l - \sum_{j=0}^{S} \frac{\tilde{\alpha}_j}{2} \alpha'_j P'_\alpha P_\alpha \alpha_j - \sum_{j=p+1}^{S} \frac{\tilde{\gamma}_j}{2} (e'_1 e_1 + 2 e'_1 P_\gamma \gamma_j + \gamma'_j P'_\gamma P_\gamma \gamma_j)$$

with $\gamma'_j = (\gamma_{2j}, \dots, \gamma_{qj})$, $e'_1 = (-1, 0, \dots, 0)$, penalty parameters $\tilde{\alpha}_j, j = 0, \dots, S$ and $\tilde{\gamma}_j, j = p + 1, \dots, S$, and penalty matrices P_α and P_γ of the form

$$P_\alpha = \begin{pmatrix} -1 & 1 & & \\ & -1 & 1 & \\ & & \ddots & \\ & & & -1 & 1 \end{pmatrix}$$

and

$$P_\gamma = \begin{pmatrix} 1 & 0 & 0 & & 0 \\ -1 & 1 & 0 & & 0 \\ 0 & -1 & 1 & & 0 \\ \vdots & & \ddots & -1 & 1 \end{pmatrix}.$$

The penalty matrices correspond to simple difference penalties, the term for the γ_js being more complicated because of the constraint $\gamma_{1j} = 1$. Higher order differences, resulting in higher order smoothness, could easily be incorporated.

The parameter vector $\delta' = (\gamma'_{p+1}, \dots, \gamma'_S, \alpha'_0, \dots, \alpha'_S)$ collects all parameters to be estimated. Typically one would proceed by solving the estimation equations $\partial l_p / \partial \delta = 0$ using the score function $s_p(\delta) = \partial l_p / \partial \delta = (\partial l_p / \partial \gamma'_{p+1}, \dots, \partial l_p / \partial \alpha'_S)'$. The problem is, that because of the attenuation parameters γ_{tj} the predictor (2.11) is non-linear and the usual (penalized) Fisher scoring can not be applied.

An approach for estimation using alternating steps is suggested: In the *first step* the parameters γ_{tj} are considered fixed. For given parameters γ_{tj} the equations

$$\frac{\partial l_p}{\partial \alpha} = 0$$

45

with $\alpha' = (\alpha'_0, \ldots, \alpha'_S)$ are solved by iterative (penalized) Fisher scoring of the form

$$\alpha^{(s+1)} = \alpha^{(s)} + F_{p,\gamma}^{-1}(\alpha^{(s)})s_{p,\gamma}(\alpha^{(s)}) \tag{2.13}$$

where the penalized score function $s_{p,\gamma}(\alpha)$ is given by $s_{p,\gamma}(\alpha) = (s'_{p,\alpha_1}, \ldots, s'_{p,\alpha_S})'$ with components

$$s_{p,\alpha_j} = \frac{\partial l_p}{\partial \alpha_j} = \sum_{t=1}^{q} \sum_{i \in R_t} \gamma_{tj} w_{itj} \frac{\partial h(\eta_{it})}{\partial \eta}(y_{it} - \mu_{it})/\sigma_{it}^2 - \tilde{\alpha}_j P'_\alpha P_\alpha \alpha_j$$

where $\mu_{it} = h(\eta_{it}), \sigma_{it}^2 = h(\eta_{it})(1 - h(\eta_{it}))$. The penalized Fisher matrix $F_{p,\gamma}(\alpha)$ is a block diagonal matrix $F_{p,\gamma}(\alpha) = Diag(F_{p,\gamma}(\alpha_0), \ldots, F_{p,\gamma}(\alpha_S))$ with components

$$F_{p,\gamma}(\alpha_j) = \sum_{t=1}^{q} \sum_{i \in R_t} \gamma_{tj}^2 w_{itj} w'_{itj} \left[\frac{\partial h(\eta_{it})}{\partial \eta}\right]^2 /\sigma_{it}^2 + \tilde{\alpha}_j P'_\alpha P_\alpha.$$

The *second step* uses the estimates of α which are resulting from the first step. If the parameters $\alpha_0, \ldots, \alpha_S$ are given, the equations

$$\frac{\partial l_p}{\partial \gamma} = 0$$

with $\gamma' = (\gamma'_{p+1}, \ldots, \gamma'_S)$ are solved by iterative (penalized) Fisher scoring of the form

$$\gamma^{(r+1)} = \gamma^{(r)} + F_\alpha^{-1}(\gamma^{(r)})s_\alpha(\gamma^{(r)}). \tag{2.14}$$

The penalized score function $s_{p,\alpha}(\gamma)$ is given by $s_{p,\alpha}(\gamma) = (s'_{p,\gamma_{1,p+1}}, \ldots, s'_{p,\gamma_{pS}})'$ with components

$$s_{p,\gamma_{tj}} = \frac{\partial l_p}{\partial \gamma_{tj}} = \sum_{i \in R_t} \alpha'_j w_{itj} \frac{\partial h(\eta_{it})}{\partial \eta}(y_{it} - \mu_{it})/\sigma_{it}^2 - \tilde{\gamma}_j\{[e_1^T P_\gamma]_t + [P'_\gamma P_\gamma]_t \gamma_j\}$$

where again $\mu_{it} = h(\eta_{it}), \sigma_{it}^2 = h(\eta_{it})(1 - h(\eta_{it}))$ and $[\ \ ]_t$ denotes the component t of a vector or, in case of a matrix, the tth row. The penalized Fisher matrix $F_{p,\alpha}(\gamma)$ is given by

$$F_\alpha(\gamma) = \sum_{t=1}^{q} \sum_{i \in R_t} v_{it} v'_{it} \left[\frac{\partial h(\eta_{it})}{\partial \eta}\right]^2 /\sigma_{it}^2 + P_{\tilde{\gamma}}$$

where $v'_{it} = (0'_q, \ldots, \tilde{w}'_{it}, \ldots, 0'_q), \tilde{w}_{it} = (\alpha'_1 w_{it1}, \ldots, \alpha'_q w_{itS})$ and 0_q is a q-dimensional vector containing zeros, i.e., $0'_q = (0, \ldots, 0)$. The penalty is given as the block diagonal matrix $P_{\tilde{\gamma}} = Diag(\tilde{\gamma}_1 P'_\gamma P_\gamma, \ldots, \tilde{\gamma}_S P'_\gamma P_\gamma)$.

The steps for estimation of α and γ are iterated until convergence. Within steps a rather coarse criterion for stopping Fisher scoring is used, which results in 3-4 repetitions of (2.13) or (2.14) per step.

Confidence bands and degrees of freedom

For illustration, and in order to judge the fit of the discrete time logistic survival model (2.10) pointwise confidence bands are wanted for the smooth covariate influence functions f_j, and the time-varying components. This allows to judge whether a simple parametric form would be sufficient, or whether there is any effect at all for a covariate. An alternative way to judge the amount of flexibility required for a covariate is the comparison of alternative models by some model selection criterion. As detailed in Section 1.3.1 of the previous chapter the effective degrees of freedom are a useful basis for such a criterion. Therefore, a hat matrix for model (2.11) will be derived. The trace of this matrix will then be used to calculate the AIC (Akaike, 1973).

For calculation of confidence bands, necessary as well as for the hat matrix, the Fisher matrix for the total parameter vector $\delta' = (\alpha', \gamma')$ is needed. It is given by $F_p(\delta) = E(-\partial^2 l_p / \partial \delta \partial \delta')$ with components

$$E\left(-\frac{\partial^2 l_p}{\partial \alpha_j \partial \alpha'_j}\right) = \sum_{t=1}^{q} \sum_{i \in R_t} \gamma_{tj}^2 w_{itj} w'_{itj} \left(\frac{\partial h(\eta_{it})}{\partial \eta}\right)^2 / \sigma_{it}^2 + \tilde{\alpha}_j P'_\alpha P_\alpha,$$

$$E\left(-\frac{\partial^2 l_p}{\partial \alpha_j \partial \alpha_l}\right) = \sum_{t=1}^{q} \sum_{i \in R_t} \gamma_{tj} \gamma_{tl} w_{itj} w'_{itl} \left(\frac{\partial h(\eta_{it})}{\partial \eta}\right)^2 / \sigma_{it}^2, l \neq j,$$

$$E\left(-\frac{\partial^2 l_p}{\partial \gamma_{tj} \partial \gamma_{tj}}\right) = \sum_{i \in R_t} (\alpha'_j w_{it})^2 \left(\frac{\partial h(\eta_{it})}{\partial \eta}\right)^2 / \sigma_{it}^2 + \tilde{\gamma}_j [P'_\gamma P_\gamma]_{tt},$$

$$E\left(-\frac{\partial^2 l_p}{\partial \gamma_{tj} \partial \gamma_{\tilde{t}j}}\right) = \tilde{\gamma}_j [P'_\gamma P_\gamma]_{t\tilde{t}} \quad , \quad \tilde{t} \neq t,$$

$$E\left(-\frac{\partial^2 l_p}{\partial \alpha_j \partial \gamma_{tj}}\right) = \sum_{i \in R_t} \gamma_{tj} w_{itj} w'_{itj} \alpha_j \left(\frac{\partial h(\eta_{it})}{\partial \eta}\right)^2 / \sigma_{it}^2,$$

and

$$E\left(-\frac{\partial^2 l_p}{\partial \alpha_j \partial \gamma_{tl}}\right) = \sum_{i \in R_t} \gamma_{tj} w_{itl} w'_{itj} \alpha_l \left(\frac{\partial h(\eta_{it})}{\partial \eta}\right)^2 / \sigma_{it}^2, l \neq j,$$

where $[\ ]_{t\tilde{t}}$ denotes the elements $(t\tilde{t})$ of a matrix.

The components of the Fisher matrix without penalty $F(\delta)$ contain the first terms as given above, i.e., with $\tilde{\alpha}_j = \tilde{\gamma}_j = 0$. Thus, one has $F_p(\delta) = F(\delta) + P$, where P is the penalty matrix depending on $\tilde{\alpha}_j, \tilde{\gamma}_j$. In the same way the score function $s(\delta) = \partial l/\partial \delta$ is related to the penalized score function $s_p(\delta) = \partial l_p/\partial \delta$ by $s_p(\delta) = s(\delta) + P\delta$. Common assumptions are $\partial l/\partial \delta = 0_p(n^{1/2})$, $\partial^2 l/\partial \delta \partial \delta' = -F + 0_p(n^{1/2})$ with $F^{-1} = 0(n^{-1})$, and there are similar assumptions for higher order derivatives. The expansion

$$0 = s_p(\hat{\delta}) = s_p(\delta) + \frac{\partial s_p}{\partial \delta'}(\hat{\delta} - \delta) + \dots$$

then yields

$$
\begin{aligned}
\hat{\delta} - \delta &= \left(-\tfrac{\partial s_p}{\partial \delta'}\right)^{-1} s_p(\delta) + 0_p(n^{-1}) + 0_p(n^{-3/2}\tilde{\lambda}) \\
&= (F(\delta) + P)^{-1}\left(s(\delta) + P\delta\right) + 0_p(n^{-1}) + 0_p(n^{-3/2}\tilde{\lambda})
\end{aligned}
$$

where $\tilde{\lambda} = \max\{\tilde{\gamma}_j, \tilde{\alpha}_j\}$. Based on this expansion the covariance may be shown to have the form

$$
\begin{aligned}
\text{cov}(\hat{\delta}) &= (F(\delta) + P)^{-1}\text{cov}(s(\delta))(F(\delta) + P)^{-1} + 0(n^{-2}) + 0(n^{-3}\lambda^2) \\
&= F_p(\delta)^{-1} F(\delta) F_p(\delta)^{-1} + 0(n^{-2}) + 0(n^{-3}\lambda^2).
\end{aligned}
$$

Thus, for calculation of pointwise confidence intervals the approximation

$$\text{cov}(\hat{\delta}) = F_p^{-1}(\hat{\delta}) F(\delta) F_p^{-1}(\hat{\delta}) \tag{2.15}$$

is used, based on asymptotic arguments when $n \to \infty$, $\tilde{\lambda}/n \to 0$, assuming that the number of knots is large enough to represent the underlying smooth structure of influential terms (see also Tutz and Scholz, 2004). For the simple case without multiplicative terms, i.e., $\gamma_{tj} = 1, t = 1, \dots, q, j = 0, \dots, S$, this approximation is identical to the sandwich matrix proposed by Marx and Eilers (1998).

To investigate how well this approximation works data are generated from model (2.20) (which will be explained in detail later) with one non-linear covariate effect that is attenuated over time. Model (2.9) was fitted to 50 data sets generated from this model with 100 and with 500 observations. Figure 2.1 shows the fitted covariate effects (left panels) and the estimated attenuation parameters (right panels). The true functions are indicated by a dotted line/circles. The mean estimate over the 50 data sets is shown by a solid line, the mean of the estimated pointwise confidence bands by dash-dotted lines and the empirical pointwise confidence bands by broken lines. For 100 observations (upper panels) the mean estimate of the covariate effect is somewhat dampened compared to the true

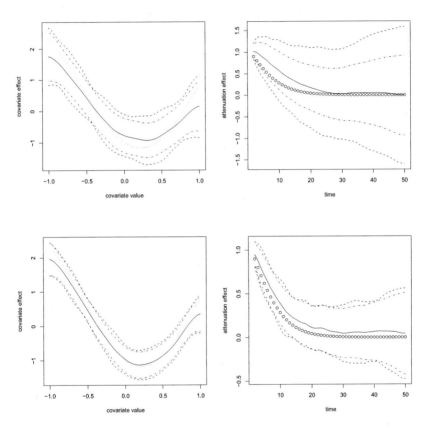

Figure 2.1: Time-dependent smooth model fitted to data generated from a model with non-linear covariate effects with attenuation over time ($c_a = 0.2$, $b_1 = 2$) for 100 (upper panels) and 500 subjects (lower panels), where the estimated smooth effect is shown in the left panels and the estimated variation with time in the right panels (true functions: dotted line/circles, mean of estimates: solid lines, empirical pointwise confidence bands: broken lines, mean of estimated pointwise confidence bands: dash-dot lines).

function. The estimated pointwise confidence bands for the covariate effect are a slightly too narrow. The estimated attenuation parameters are seen to lag considerably behind the true parameters. The empirical as well the estimated pointwise confidence bands for the attenuation parameters are very wide, where the latter are too narrow. For 500 observations (lower panels) the true covariate effect as well as the true attenuation parameters are approximated though very well. Also the (mean) estimated pointwise confidence bands are close to the empirical bands, i.e., the covariance approximation appears to work properly.

By treating the Fisher matrix $F(\delta)$ introduced above as if it would be from a *one step* iterative penalized estimation procedure for the parameter vector δ the hat matrix H is obtained by $H = (F(\hat{\delta}) + P)^{-1} F(\hat{\delta})$ (Eilers and Marx, 1996). Its trace $tr(H)$ can be taken for the effective degrees of freedom of a model (Hastie and Tibshirani, 1990). This measure of model complexity can be used for calculation of various model selection criteria. For example the AIC is obtained by

$$\text{AIC} = -2 \cdot l_p + 2 \cdot \text{tr}(H). \tag{2.16}$$

In the following this criterion will be used for selection of penalization parameters as well as for comparing several candidate models for deciding which one is the most appropriate.

2.2 Simulation study

In the previous section a flexible model for discrete time survival data (2.10) was introduced that allows for non-linear covariate influence as well as for time-varying effects. It encompasses four sub-models of varying complexity: a simple parametric model (2.5), a time-dependent model (2.6), a "smooth" model with non-linear covariate effect (2.8) and a time-dependent smooth model (2.9). In the following these models are fitted to simulated data. The advantage of such a simulation study is that the true structure is known and therefore by systematical variation the properties of the fitted models can be explored directly.

The example data are generated from models of the type (2.5) - (2.9). There are four categories of model fit (illustrated in Figure 2.2) that result from the type of flexibility required by such data in combination with the flexibility provided by the model used for fitting. When the *adequate model* is used for fitting, i.e., just the required amount of flexibility is provided, the sole problems expected are due to an insufficient amount of information, i.e., a small number of subjects or

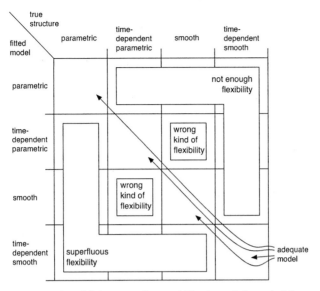

Figure 2.2: Categories of model fit arising from combinations of the underlying true structure (columns) and the model used for fitting (rows).

a small signal-to-noise ratio, or (closely connected) due to an inadequate choice of the penalty parameters. Especially the role of the penalty parameter is important when the model used for fitting provides *superfluous flexibility*. In case of superfluous flexibility criteria such as the AIC ideally should indicate that a simpler model is adequate, or selection of the penalty parameters should be guided to provide flexibility where needed only. If *not enough flexibility* is provided by the model used for fitting, the shape of the resulting fit as well as the loss in performance compared to the adequate model are going to be investigated. This loss in performance will also be compared to that resulting from models that provide superfluous flexibility. This will help to answer the question whether in cases of uncertainty about the true structure simpler models (that might be too simple) or more flexible models (that might be too flexible) should be chosen. For models that provide the *wrong kind of flexibilty*, i.e., flexibility where it is not needed and not enough flexibility where it should be provided, the fits are going to be examined with respect to their performance as well as with respect to their shape.

To keep the structure of the data simple examples with only one covariate $x_i, i =$

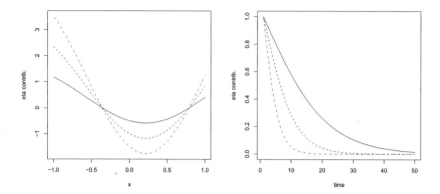

Figure 2.3: Non-linear covariate effect of the simulated data (left panel) for $b_1 = 1$ (solid line) $b_1 = 2$ (broken line) and $b_1 = 3$ (dash-dotted line) and attenuation effect (right panel) for $c_a = 0.1$ (solid line), $c_a = 0.2$ (broken line) and $c_a = 0.4$ (dash-dotted line)

$1, \ldots, n$ drawn from a uniform distribution (with values between -1 and 1) are used. In addition the generating models have a fixed baseline. The generating model corresponding to model (2.5) is one with a linear covariate effect and no variation with time

$$\eta_{it} = b_0 + b_1 x \qquad (2.17)$$

with $b_0 = \log(\frac{0.05}{1-0.05})$ (i.e., the baseline hazard is 0.05) and covariate influence parameter b_1. The generating model used for a non-linear covariate effect (corresponding to (2.8)) is

$$\eta_{it} = b_0 + b_1 \cdot \sin(2(x_i - 1)) \qquad (2.18)$$

where b_0 is chosen such that the baseline hazard is 0.05 ($b_0 = \log(\frac{0.05}{1-0.05}) - \int_{-1}^{1} b_1 \sin(2(x-1))dx$), and b_1 determines the amount of covariate influence. The left panel of Figure 2.3 shows the non-linear component $b_1 \cdot \sin(2(x_i - 1))$ for several values b_1. Its shape was chosen such that a linear approximation can cover some aspects of it when a line with negative slope is fitted, but important features (the increase for large values of x) are missed. For a linear covariate influence that diminishes over time corresponding to model (2.6) the generating model

$$\eta_{it} = b_0 + \frac{2}{1 + \exp(c_a(t-1))} b_1 x_i \qquad (2.19)$$

is used, where again the parameter b_0 is chosen such that the baseline hazard is 0.05, and b_1 determines the amount of covariate influence. The parameter c_a

determines how fast the covariate influence decreases towards zero over time. The right panel of Figure 2.3 shows the effect of the component $\frac{2}{1+\exp(c(t-1))}$ for several values c_a. Combining the non-linear effect from (2.18) and the time-varying effect from (2.19) a generating model with time-varying non-linear covariate influence corresponding to model (2.9) is obtained in the form

$$\eta_{it} = b_0 + \frac{2}{1 + \exp(c_a(t - 1))} b_1 \cdot \sin(2(x_i - 1)) \tag{2.20}$$

with b_0 chosen such that the baseline hazard is 0.05.

The observed time t_i is generated by drawing pseudo-responses y_{it} for each subject i successively for each time $t = 1, \ldots, 50$ from a binomial distribution $y_{it} \sim B(1, \frac{\exp(\eta_{it})}{1+\exp(\eta_{it})})$ until $y_{it} = 1$ is obtained, or $t = 50$. The maximum value reached for time t is taken to be the observed time t_i. If $t_i = 50$ the indicator δ_i is set to $\delta_i = 0$, taking the value $\delta_i = 1$ otherwise. This simple censoring scheme is chosen to keep the true structure simple.

Figure 2.4 shows the *survival function*

$$S(t|x) = P(T > t|x) = \prod_{i=1}^{t}(1 - \lambda(i|x))$$

for several types of subjects. The survival function corresponding to the baseline hazard 0.05 is indicated by circles and the survival functions for subjects with covariate contribution $+2$ (this obtained for example from model (2.19) with $b_1 = 2, c_a = 0$, and covariate value $x_i = 1$) and -2 are indicated by triangles. The survival functions that are obtained by attenuating such covariate effects over time (with $c_a = 0.2$) are indicated by crosses. It can be seen that starting at about $t = 10$ the shape of these survival functions approaches the one of the baseline survival function.

Besides using different generating models with various amounts of covariate influence (with larger values of b_1 resulting in an increased signal-to-noise ratio) and attenuation, effects of the number of subjects are explored. Model fits are examined for $n = 50$ and $n = 100$. Whereas from the example shown in Figure 2.1 it seems to be problematic to use flexible models with a small number of subjects, it will be interesting to see if — nevertheless — a performance benefit can be obtained. The performance of the fitted models is evaluated by mean squared error

$$MSE = \frac{1}{n} \sum_{i=1}^{n} \sum_{t=1}^{T_{eval}} (p_{it} - \hat{p}_{it})^2 \tag{2.21}$$

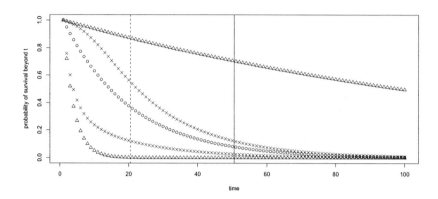

Figure 2.4: Survival function for baseline hazard 0.05 (circles), subjects with covariate contribution +2 and -2 (triangles) and subjects with covariate contribution +2 and -2 with attenuation $c_a = 0.2$ (crosses). The vertical solid line indicates the censoring used for the simulated data and the vertical broken line the upper bound of the interval used for evaluation of performance.

using a fixed value T_{eval}. p_{it} denotes the probability corresponding to a multinomial distribution with $\sum_t p_{it} = 1$. With $\lambda_{it} = \lambda(t|x_i) = \frac{\exp(\eta_{it})}{1+\exp(\eta_{it})}$ denoting the hazard function from the generating model, it is obtained by $p_{it} = \lambda_{it} \prod_{s=1}^{t-1}(1-\lambda_{is})$ for $t < T_{eval}$, and $p_{it} = \prod_{s=1}^{t-1}(1-\lambda_{is})$ for $t = T_{eval}$. The estimates $\hat{p}_{it}$ from models (2.5) - (2.9) are given in the same way with λ_{it} being replaced by $\hat{\lambda}_{it} = \frac{\exp(\hat{\eta}_{it})}{1+\exp(\hat{\eta}_{it})}$. In the following the value $T_{eval} = 20$ is used, for differences in fitting of time-varying structure are mainly found in the interval [1,20] (see e.g. Figure 2.4). Using a larger interval would give less weight to the early observation intervals and therefore obscure performance differences resulting from adequate fitting of attenuation structure. For each example the mean squared errors for 50 repetitions of data generation and model fitting are reported.

For fitting model (2.5) to data with one covariate, one penalty parameter (for the baseline) has to be selected, for models (2.6) and (2.8) two penalty parameters have to be set, and model (2.9) requires three penalty parameters. AIC is evaluated (2.16) for 10 values, a 10×10 grid and a $10 \times 10 \times 10$ grid respectively and that penalty parameters are chosen that result in the best value.

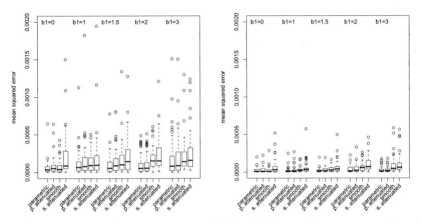

Figure 2.5: Mean squared errors for various models fitted to data with linear covariate effect and no variation over for various amounts of covariate influence b_1 for 50 (left panel) and 100 subjects (right panel).

2.2.1 Parametric and non-linear effects without time-variation

Figure 2.5 shows the mean squared errors for data with linear covariate effect, and no effect of time generated from model (2.5) with various amounts of covariate effect ($b_1 \in \{0, 1, 1.5, 2, 3\}$) for 50 (left panel) and 100 subjects (right panel). When there is no effect of the covariate ($b_1 = 0$) all models perform similar, with the exception of the most flexible model (2.9), which allows for a time-dependent smooth covariate effect. The latter though performs distinctly worse, indicating that the model provides too much flexibility that is not regulated appropriately. For the other models restriction of flexibility seems to be effective. When there is a linear covariate effect in the data ($b_1 \neq 0$) that does not vary with time, the simplest model (2.5) performs best, and performance drops when the amount of flexibility provided by the fitted model increases. In some examples (e.g. $n = 50, b_1 = 2$ or $n = 100, b_1 = 3$) the models (2.8) and (2.9) that allow for a non-linear covariate effect perform poorly while models allowing for a time-varying linear effect incur a much smaller loss of performance as compared to the simple parametric model. The mean squared errors for $n = 100$ are can be seen to be considerably smaller than those for $n = 50$.

The AIC, based on the effective degrees of freedom, has been used to select the

Table 2.1: Proportion of repetitions where a specific type of model (pa.: parametric covariate effect; pa.a.: parametric attenuated; sm.: smooth covariate effect; sm.a.: smooth attenuated) would have been selected by AIC for data generated from a model with a linear or non-linear covariate effect that does not change over time. The largest proportion for each example is indicated in boldface.

b_1	pa.	pa.a.	sm.	sm.a.	pa.	pa.a.	sm.	sm.a.
		$n = 50$				$n = 100$		
linear covariate effect								
0	0.04	0.04	0.26	**0.66**	0.12	0.08	0.26	**0.54**
1	0.28	0.18	0.18	**0.36**	**0.48**	0.10	0.12	0.30
1.5	**0.46**	0.18	0.06	0.30	**0.46**	0.26	0.08	0.20
2	**0.50**	0.24	0.08	0.18	**0.64**	0.12	0.06	0.18
3	**0.48**	0.26	0.04	0.22	**0.34**	0.30	0.20	0.16
non-linear covariate effect								
1	0.04	0.02	0.32	**0.62**	0	0	0.42	**0.58**
1.5	0	0.06	0.34	**0.60**	0	0	**0.50**	**0.50**
2	0	0.02	0.46	**0.52**	0	0	**0.60**	0.40
3	0	0	0.46	**0.54**	0	0	**0.60**	0.40

penalty parameters for the model fits shown in Figure 2.5. It can also be used to compare alternative models. The upper part of Table 2.1 indicates for each example of Figure 2.5 which model would have been chosen by AIC for what proportion of repetitions. For most of the examples the adequate model (2.5) has the largest share. Exceptions are seen for examples with with a low amount of information ($b_1 = 0$ and $b_1 = 1$ for $n = 50$ and $b_1 = 0$ for $n = 100$), where the most flexible model would be selected, resulting in a very bad performance.

Figure 2.6 shows the results for data generated from model (2.18) with a non-linear covariate effect, and without an effect of time. The adequate model now is model (2.8) which allows for a smooth function for the covariate effect. Comparing its performance to that of model (2.5), which only allows for a linear covariate effect, it is obvious that there is a distinct benefit of using a model with enough flexibility. Only when the covariate effect is very weak, and the number of subjects is small ($n = 50$, $b_1 = 50$), there is no more advantage of using models allowing for smooth covariate effects. For all other examples also model (2.9), which allows for variation of a smooth covariate effect over time and therefore provides too much flexibility, outperforms the parametric models. This indicates that control of flexibility by AIC grid search works satisfyingly. The lower part of Table 2.1 indicates that AIC is suitable also for model selection, as it shows for what proportion of repetitions which model would have been selected. For $n = 100$ for every repetition a model would have been selected that does provide enough flexibility. For some of the repetitions of the examples with $n = 50$ parametric

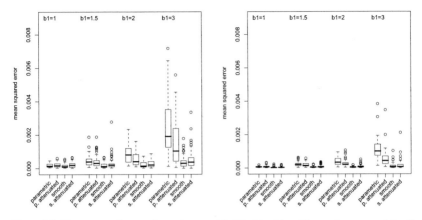

Figure 2.6: Mean squared errors for various models fitted to data with non-linear covariate effect and no variation over time for various amounts of covariate influence b_1 for 50 (left panel) and 100 subjects (right panel).

models would be selected by AIC. As furthermore the mean squared errors for $n = 50$ are larger than those for $n = 100$ the selection of parametric models might be due to an insufficient amount of information in the data that does not allow for adequate estimation of the non-linear effect.

Model (2.6), which allows for variation over time instead of allowing for smooth covariate effects, provides the wrong kind of flexibility for the data with non-linear covariate effect and no effect of time. Nevertheless, for $b_1 \geq 1.5$ the resulting mean squared errors in Figure 2.6 are smaller than those for the parametric model (2.5), which does not allow for effects of time. Examining the model fits should provide some hints about the source of this performance difference. Figure 2.7 shows the mean estimated baseline (left panel) and the mean estimated covariate effect (right panel) for one example averaged over the 50 repetitions. It can be seen that although the generating model uses a constant baseline and there is no variation over time, the fits indicate a declining baseline and a covariate effect with a negative slope that is attenuated with time. Considering how a linear term would be fitted to the non-linear true function (left panel of Figure 2.3), it seems plausible that in the beginning, with all subjects present, a negative slope provides the closest approximation. After subjects with large positive covariate contribution have dropped out a smaller slope yields a better approximation. Therefore the model fit indicates variation with time. As in addition the resulting performance is better than that of the simple parametric model (2.5) one risks to

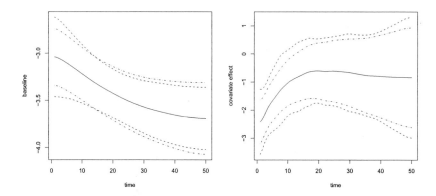

Figure 2.7: Time-dependent parametric model fitted to data generated from a model with non-linear covariate effects with no variation over time ($n = 100$, $c_a = 0$, $b_1 = 3$) where the estimated baseline is shown in the left panel and the estimated covariate effect is shown in the right panel (mean of estimates: solid line; empirical pointwise confidence bands: broken lines; mean of estimated pointwise confidence bands: dash-dotted lines).

be misguided with respect to the underlying true structure. This can be avoided by consulting the AIC. As already seen in Table 2.1 adequate models are identified by this criterion in almost all repetitions.

2.2.2 Parametric and non-linear effects with time-variation

In the following generating models (2.19) and (2.20) are used that incorporate covariate effects that are attenuated over time. Figure 2.8 shows the mean squared errors for model fits to data generated from model (2.19) with an attenuated linear covariate effect. It can be seen that the adequate model (2.6), which allows for linear covariate effect that varies over time, has the best performance for all examples. When attenuation in the generating model is small and the covariate effect itself also is small ($c_a = 0.1$, $b_1 = 1$; upper panels) the parametric model (2.5), that does not allow for an effect of time, performs similar. For $b_1 \geq$ 1.5 model (2.9) with time-varying smooth covariate effect is a close contender, indicating that the superfluous flexibility is controlled adequately by selection of penalties by AIC. The upper part of Table 2.2 shows which models would

58

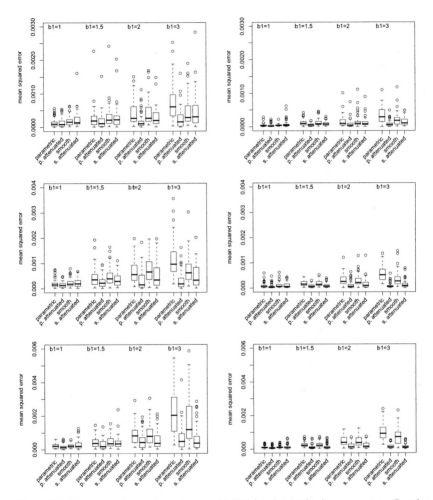

Figure 2.8: Mean squared errors for various models fitted to data with a linear covariate effect that is attenuated over time for various amounts of covariate influence b_1) and attenuation parameters $c_a = 0.1$ (upper panels), $c_a = 0.2$ (middle panels) and $c_a = 0.4$ (lower panels) and 50 (left panels) or 100 subjects (right panels).

Table 2.2: Proportion of repetitions where a specific type of model (pa.: parametric covariate effect; pa.a.: parametric attenuated; sm.: smooth covariate effect; sm.a.: smooth attenuated) would have been selected by AIC for data generated from a model with a linear or non-linear covariate effect that is attenuated over time. The largest proportion for each example is indicated in boldface.

c_a	b_1	pa.	pa.a.	sm.	sm.a.	pa.	pa.a.	sm.	sm.a.
			$n = 50$				$n = 100$		
linear covariate effect									
0.1	1	0.10	0.22	0.22	**0.46**	0.16	0.30	0.18	**0.36**
	1.5	0.16	**0.40**	0.04	**0.40**	0.02	**0.48**	0.16	0.34
	2	0.06	**0.50**	0.10	0.34	0.04	**0.60**	0.12	0.24
	3	0.02	**0.70**	0.06	0.22	0.00	**0.78**	0.02	0.20
0.2	1	0.08	0.26	0.16	**0.50**	0.02	0.22	0.20	**0.56**
	1.5	0.02	0.30	0.12	**0.56**	0.02	**0.56**	0.16	0.26
	2	0.06	**0.48**	0.06	0.40	0.02	**0.54**	0.08	0.36
	3	0.04	**0.52**	0.16	0.28	0.00	**0.62**	0.00	0.38
0.4	1	0.08	0.08	0.24	**0.60**	0.04	0.12	0.22	**0.62**
	1.5	0.08	0.18	0.24	**0.50**	0.06	0.26	0.12	**0.56**
	2	0.08	0.20	0.14	**0.58**	0.06	**0.44**	0.06	**0.44**
	3	0.00	0.48	0.02	**0.50**	0.00	**0.54**	0.02	0.44
non-linear covariate effect									
0.1	1	0.06	0.16	0.26	**0.52**	0.02	0.08	0.30	**0.60**
	1.5	0.04	0.04	0.26	**0.66**	0.00	0.12	0.16	**0.72**
	2	0.00	0.10	0.20	**0.70**	0.00	0.02	0.20	**0.78**
	3	0.00	0.02	0.24	**0.74**	0.00	0.00	0.08	**0.92**
0.2	1	0.06	0.08	0.28	**0.58**	0.06	0.20	0.22	**0.52**
	1.5	0.00	0.02	0.20	**0.78**	0.00	0.02	0.16	**0.82**
	2	0.04	0.10	0.14	**0.72**	0.00	0.02	0.06	**0.92**
	3	0.00	0.04	0.24	**0.72**	0.00	0.00	0.12	**0.88**
0.4	1	0.10	0.04	0.28	**0.58**	0.06	0.10	0.14	**0.70**
	1.5	0.00	0.10	0.16	**0.74**	0.02	0.10	0.08	**0.80**
	2	0.02	0.14	0.20	**0.64**	0.02	0.14	0.12	**0.72**
	3	0.00	0.04	0.14	**0.82**	0.00	0.04	0.04	**0.92**

be selected when AIC is used not only for selection of penalty parameters but also for selecting models. It can be seen that for the majority of repetitions models get selected that provide a sufficient amount of flexibility. Several times model (2.9) is picked, which actually provides too much flexibility, but as already mentioned performs similar to the adequate model (2.6) due to appropriately selected penalty parameters. Despite less information in the data, indicated by a generally larger mean squared error, this is even valid for $n = 50$ (for $b \geq 1.5$), i.e., when only few subjects are available.

For the data with an attenuated parametric covariate effect, model (2.8) provides

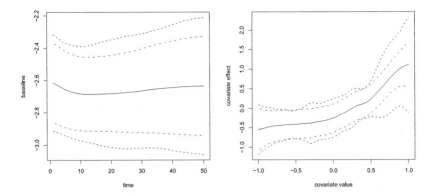

Figure 2.9: Estimated smooth effect of a model that does not allow for variation with time fitted to linear covariate effect data with time attenuation ($n = 100$, $c_a = 0.4$, $b_1 = 3$) where the estimated baseline is shown in the left panel and the estimated smooth covariate effect in the right panel (mean of estimates: solid line, empirical pointwise confidence bands: broken lines, mean of estimated pointwise confidence bands: dash-dot lines).

the wrong kind of flexibility by allowing for a smooth covariate effect instead of allowing for an effect of time. It can be seen from Figure 2.8 that, while not being competitive to models (2.6) and (2.9), in some examples (e.g. for $b_1 = 3$) it performs slightly better than the simple parametric model (2.5). Similar to the "wrong kind of flexibility" example in Section 2.2.1 therefore the model fit is examined. Figure 2.9 shows the estimates for one of the examples averaged over the 50 repetitions. It can be seen that a slightly non-linear effect (with increasing slope for large covariate values) is fitted, which is not present in the generating model. This again is an obvious example where false conclusions about the underlying true structure can be avoided by inspecting the AIC (see Table 2.2).

Figure 2.10 shows the results for data generated from model (2.20) that incorporates a non-linear covariate effect that becomes attenuated over time. The adequate model (2.9) is shown to perform best generally, except for examples with small covariate effect ($b_1 = 1$). All other models considered here do not provide enough flexibility. Model (2.8) which allows for a smooth covariate effect but not for an effect of time shows competitive performance for a small number of subjects ($n = 50$) and up to a moderate amount of attenuation ($c_a \leq 0.2$). This indicates that a certain amount of information is needed for a fit of model

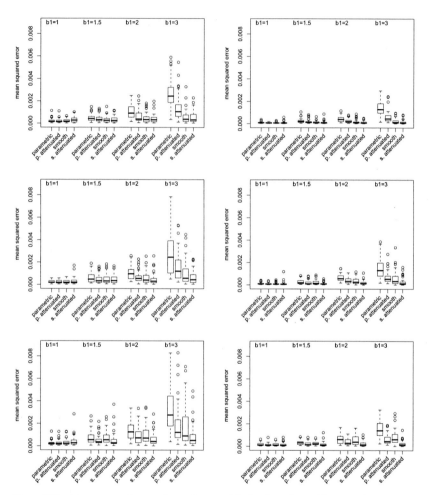

Figure 2.10: Mean squared errors for various models fitted to data with a non-linear covariate effect that is attenuated over time with various amounts of covariate influence b_1 and attenuation parameters $c_a = 0.1$ (upper panels), $c_a = 0.2$ (middle panels) and $c_a = 0.4$ (lower panels) and 50 (left panels) or 100 subjects (right panels).

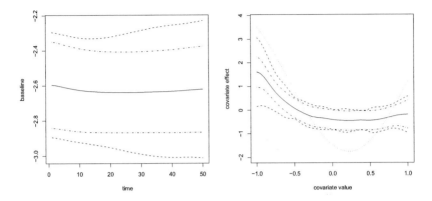

Figure 2.11: Estimated smooth effect of a model that does not allow for variation over time fitted to a non-linear covariate effect data with attenuation over time ($c_a = 0.4$, $b_1 = 3$) where the left panel shows the estimated baseline and the right panel the estimated covariate effect (mean of estimates: solid line, empirical pointwise confidence bands: broken lines, mean of estimated pointwise confidence bands: dash-dot lines, true function: dotted line).

(2.9) that uncovers all of the underlying structure. Model (2.6) which allows for a time-dependent linear covariate effect is shown to perform slightly better than the simple parametric model (2.5). Despite the inadequate shape of the covariate influence assumed by model (2.6) still the underlying time-varying structure of the data seems to be detected. The lower part of Table 2.2 indicates which models would have been selected based on AIC. It can be seen that for most of the examples and for most of the repetitions the adequate model (2.9) is selected. Only for a small to medium attenuation, the smooth model (2.8) which does not allow for an effect of time has a somewhat larger share. On the other hand its performance can be seen to be competitive in Figure 2.10 for these examples with small to medium attenuation.

To illustrate the effect of fitting a model that does not allow for variation of the covariate effect over time, Figure 2.11 shows the mean of the estimates from model (2.8) for one of the examples from Figure 2.10. It can be seen that the estimate is a dampened version of the true function and that it does not share all of its features (e.g. the increase for large covariate values). A possible explanation would be that the smooth function is estimated at an intermediate level of attenuation.

2.3 Real data

The data presented here are from 1922 patients of a German psychiatric hospital. The response is time spent in hospital, measured in days. The covariates available for smooth modelling are age at admission, calendar time measured in days for a period of five years, and GAF (Global Assessment of Functioning) score at admission which is a physician's judgment of the patients' level of functioning. The other covariates are 0-1-coded variables and are given by: gender (MALE=1: male), education (EDU=1: above high school level), partner situation (PART=1: has a permanent partner), job situation (JOB=1: full/part time job at admission), first hospitalization (FIRST=1: first admission in a psychiatric hospital) and suicidal action (SUI=1: suicidal act previous to admission).

The predictor of the model for "probability of discharge" has the form

$$
\begin{aligned}
\eta_{it} = \ & \beta_{t0} + f_T(\text{calendar time}) + \\
& \text{MALE} \cdot \beta_{\text{MALE}} + \text{EDU} \cdot \beta_{\text{EDU}} + \text{PART} \cdot \beta_{\text{PART}} + \\
& \text{JOB} \cdot \beta_{\text{JOB}} + \text{FIRST} \cdot \beta_{\text{FIRST}} + \text{SUI} \cdot \beta_{\text{SUI}} + \\
& f_A(\text{age}) + \gamma_t \cdot f_G(\text{GAF score})
\end{aligned}
$$

where f_T, f_A and f_G are smooth functions. A specific feature of the model is the incorporation of two time scales, namely the time patients stay in the hospital, and the calendar time. The flexibility of the model allows for modelling of both effects.

The predictor given is already the result of model selection: Initially all covariates (except "calendar time") were allowed to be estimated with time-varying effects and all metric covariates were included as components with smoothly transformed effects. This general model was simplified subsequently by excluding time-varying effects for certain covariates and replacing smooth components by parametric ones. Based on AIC the model presented above was selected. Table 2.3 gives the AIC for this model and some simpler variants. It can be seen that no further simplifications are indicated. In particular the AIC favors the inclusion of a time-varying smooth term for "GAF score at admission" over a (time-dependent) parametric or a fixed smooth term.

Table 2.4 shows the parameter estimates of the parametric terms. The only variable that seems to have an influence is partner situation. The effect for this variable is rather strong. If the patient has a permanent partner the time spent in the hospital is clearly reduced. All remaining variables have no significant effect.

Table 2.3: AIC for different models for "probability of discharge" with optimal penalty parameters ("…" refers to the binary components mentioned in the text, "ct" to calendar time, "age" to age at admission and "GAF" to GAF score at admission).

Model	AIC
$\eta_{it} = \beta_{t0} + \text{ct} \cdot \beta_T + \ldots + \text{age} \cdot \beta_A + \text{GAF} \cdot \beta_G$	15790.7
$\eta_{it} = \beta_{t0} + \text{ct} \cdot \beta_T + \ldots + \text{age} \cdot \beta_A + \gamma_t \cdot f_G(\text{GAF})$	15775.3
$\eta_{it} = \beta_{t0} + f_T(\text{ct}) + \ldots + \text{age} \cdot \beta_A + \gamma_t \cdot f_G(\text{GAF})$	15773.8
$\eta_{it} = \beta_{t0} + f_T(\text{ct}) + \ldots + f_A(\text{age}) + f_G(\text{GAF})$	15777.1
$\eta_{it} = \beta_{t0} + f_T(\text{ct}) + \ldots + f_A(\text{age}) + \text{GAF} \cdot \beta_{G_t}$	15772.6
$\eta_{it} = \beta_{t0} + f_T(\text{ct}) + \ldots + f_A(\text{age}) + \gamma_t \cdot f_G(\text{GAF})$	15770.3

Table 2.4: Estimates of the parametric terms of the model for "probability of discharge"

covariate	parameter estimate	standard deviation
gender (male)	-0.0525	0.0550
education	-0.0387	0.0716
partner situation	0.2619	0.0654
job situation	0.0428	0.0731
first hospitalization	0.0845	0.0680
suicidal action	-0.0828	0.1145

The effect of calendar time is shown in the left panel of Figure 2.12. It can be seen that the time spent in hospital decreases almost continuously with calendar time. However it is not clear if the effect is more connected to the organization of the hospital or to new developments in the treatment. The right panel of Figure 2.12 shows the smooth estimate of the effect of the variable age at admission. With increasing age the stay in the hospital is shortened.

An interesting effect is that of the GAF score, which is an assessment score at admission. Figure 2.13 shows the estimated effect f_G in the left panel and the corresponding time-varying attenuation parameters γ_t in the right panel. The shape of the smooth function clearly indicates that a lower GAF score at admission (indicating a lower level of functioning) results in a lower probability of discharge. This is reasonable as a patient in a worse condition can be expected to require longer treatment. Especially the plateaus at scores under 30 and over 60 deserve attention though. They indicate that the essential difference is between low and high GAF score. Only for a small window between 30 and 50 points the effect is changing. Moreover, the curve clearly shows that a linear effect with time-varying coefficients is not appropriate. This is also supported by the AIC which favors the time-dependent smooth component for the covariate "GAF score at admission" over a time-dependent parametric component. The estimate of γ_t (right panel of Figure 2.13) shows that the effect of the GAF score at admission

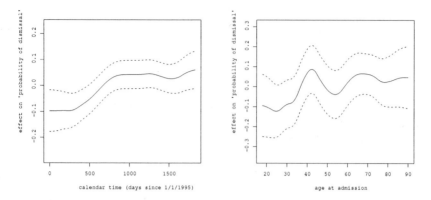

Figure 2.12: Estimated smooth effect of calendar time f_T (left panel) and smooth effect of age at admission f_A (right panel) of the model for "probability of discharge" (estimates: solid lines; pointwise confidence bands: dashed lines).

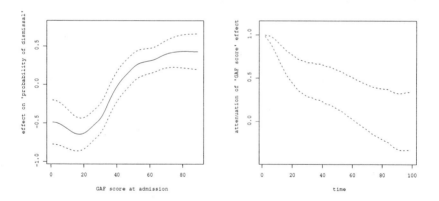

Figure 2.13: Estimated smooth effect of the covariate "GAF score at admission" on "probability of discharge" (left panel) and its attenuation over time (right panel) (estimate: solid line/circles; pointwise confidence bands: dashed lines).

vanishes over time. This is reasonable as the condition of a patient is expected to change over time, and so the predictive power of the initial score diminishes.

2.4 Discussion

In the preceding sections a new class of flexible discrete time logistic survival models was introduced that allow for smooth effects of covariates that vary with time. Even for a rather small number of subjects (as few as 50) distinct performance benefits can be expected when the underlying structure of the data to be fitted requires such flexibility. The AIC, based on effective degrees of freedom, has been demonstrated to work well for selection of the penalty parameters and for identification of adequate models. Therefore the models can be applied to data at hand with a very low risk of overfitting.

Problems might occur if there is a large number of covariates, and therefore a large number of penalty parameters needs to be selected, leading to large computational complexity. In combination with a small number of subjects, estimation might not even succeed due to convergence problems for some very flexible candidate models. A potential solution to this problem could be based on a varying-coefficient extension of the boosting approach introduced in the previous chapter.

Chapter 3

Localized logistic regression

In the previous two chapters new models and new techniques for estimation of model parameters were developed which resulted, besides prediction performance, in interpretable model fits. One of the assumptions that was employed to warrant interpretability was that of additivity of covariate effects. In the following the focus will be on applications where prediction performance is the main objective. Therefore the assumption of additivity will be dropped. Specifically data will be analyzed where the response variable is a 0/1 indicator of class membership for two classes. The corresponding technique within the generalized linear model framework (Nelder and Wedderburn, 1972; McCullagh and Nelder, 1989) that is going to be used as basis for model building is logistic regression. It is closely related to linear discriminant analysis (Fisher, 1936, 1938), which is an early attempt to deal with simple class structure by linear class boundaries. An alternative early technique that can be used with arbitrarily complex class boundaries is the nearest neighbourhood classification procedure (Fix and Hodges, 1951). It determines the class of a new observation by the majority of the class of nearby observations, therefore implicitly estimating class densities. Due to the flexibility of this approach its use is very problematic as the number of covariates increases, because the neighbourhoods become very large. This problem, in a more general form, has been termed *curse of dimensionality* (Bellman, 1961). Therefore the aim is for modern classification procedures to provide enough flexibility for the classification problems at hand while alleviating the curse of dimensionality by appropriate control of this flexibility. Hastie et al. (2001) and Duda et al. (2001) give a comprehensive overview of such modern techniques.

The classification problem given in the left panel of Figure 3.1 motivates the procedure that is going to be developed in the following. There are two informative

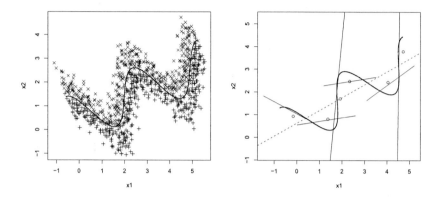

Figure 3.1: Data with two classes (crosses and plusses) with a moderately complex smooth class boundary (solid curve) in the left panel and local fitting (solid lines) guided by clusters of misclassified points (k-means centers: circles) from an initially fitted linear class boundary (broken line) in the right panel.

covariates with a moderately complex smooth optimal class boundary (solid line). Simple logistic regression leads to a linear approximation (broken line in the right panel) which is clearly not appropriate and will result in a large misclassification rate for new data. Using the boosting approach (Schapire, 1990; Freund and Schapire, 1996; Friedman et al., 2000) reviewed in Chapter 1, the initial fit could be improved by enhancing it with additional model fits, for which observations that were initially misclassified receive larger weight. While the initial fit provided a global model, clustering of the misclassified observations (obtained by k-means clustering, with its centers indicated by circles in the right panel of Figure 3.1) indicates that it might be beneficial to fit additional models locally. The solid lines in the right panel of Figure 3.1 indicate such model fits which are local with respect to clusters of misclassified observations. Alternatively, if models are fitted locally with respect to observations for which prediction is wanted, another variant of local logistic regression is obtained. Furthermore there exist approaches where models are fitted locally with respect to distance from the class boundary (Hand and Vinciotti, 2003). In the following an approach is used where models are fitted locally with respect to new observations. This has the benefit that no initial fit of a global model is required and therefore the local fits do not depend on the quality of such a global fit.

Local fitting of binary regression models has been investigated carefully by Fan

70

and Gijbels (1996) and Loader (1999). Loader (1999, Chapter 8) also investigated the use in discrimination, but only for a small number of covariates (with a maximum of four). For a larger number of covariates the curse of dimensionality becomes more severe. If the same number of observations is used for local model building then estimates are hardly local anymore. Instead of placing restrictions on the data structure, e.g. by requiring additivity of covariate influence (and therefore reducing the number of observations needed for stable estimates), local dimension reduction will be used to address this issue. A strong tool for dimension reduction is the selection of relevant covariates. Even for global classifiers performance often improves if only a small subset of informative covariates is used instead of the whole set of covariates. It is to be assumed that this idea also works locally. Moreover, for different observations different covariates may carry the relevant information. Thus in the following dimension reduction is obtained by locally adaptive selection of covariates. For alternative approaches of local dimension reduction see Schaal et al. (1998), Hastie and Tibshirani (1996). In contrast to these approaches variable selection has the advantage that one obtains information about the relevance of covariates even if the selection is performed locally.

In Section 3.1 the idea of localized logistic regression is formalized and an algorithm is introduced. For its components several options are considered. The new procedure and its variants are compared to existing procedures using simulated data in Section 3.2 and using real data in Section 3.3. In Section 3.4 a final discussion of the new technique is provided and recommendations for its application are given.

3.1   Theory

Let (x_i, y_i), $i = 1, \ldots, n_L$, denote the training set with $x'_i = (x_{i1}, \ldots, x_{ip})$ being measurements on p covariates and $y_i \in \{0, 1\}$ representing class membership. The objective is to predict the class membership of a new observation with measurement x by using x and the information from the training set.

Successful use of localization and selection of covariates depends on tuning parameters which have to be chosen. These parameters, which for example determine the amount of localization and thresholds for selection of covariates, are introduced in the following. They are considered as flexible parameters that are chosen data-adaptively. In addition there are design decisions such as which type of kernel is used for calculation of local weights.

3.1.1 Localized logistic regression

The parametric model that is localized is the well known logistic regression model

$$\log\left\{\frac{P(y_i = 1|x_i)}{1 - P(y_i = 1|x_i)}\right\} = z_i'\beta$$

where β is a parameter vector of length $m + 1$ and z_i is a design vector built from x_i. For linear logistic discrimination $z_i' = (1, x_i')$ and for quadratic logistic discrimination $z_i' = (1, x_i', x_{i1}^2, \ldots, x_{ip}^2)$ is used. The design vector could even be more complex containing for example cubic components x_{ij}^3 or interaction terms, but as its size increases estimation becomes more difficult. Later a technique will be introduced that makes it feasible to estimate model parameters even with very few observations available, if the design vector and therefore the parameter vector β is of reasonable size. Especially interaction terms of the form $x_{ij}x_{kl}$, whose number increases dramatically with increasing p, are deliberately left out to keep the number of parameters in β small. The interaction structure of several covariates can alternatively be modeled by using local models in which the contribution of each covariate varies depending on the values of the other covariates. Note that in the following the elements of z_i will be called *predictors*, in contrast to the *covariates* in x_i.

Local versions of the model are obtained by introducing weights into the (log-)likelihood (see e.g. Loader, 1999; Fan and Gijbels, 1996). For target value x the weighted log-likelihood is given by

$$l_x(\beta) = \sum_i \left(y_i \log \pi(x_i) + (1 - y_i) \log(1 - \pi(x_i))\right) w_k(z, z_i) \tag{3.1}$$

where $\pi(x_i) = P(y_i = 1|x_i)$ and z, z_i are the predictor values connected to x, x_i, i.e., $z' = (1, x')$, $z_i = (1, x_i')$ in the linear case, and $z' = (1, x', x_1^2, \ldots, x_p^2)$, $z_i' = (1, x_i', x_{i1}^2, \ldots, x_{ip}^2)$ in the quadratic case. The locally adaptive weights $w_k(z, z_i)$ are chosen to depend on the (Euclidian) distance between the (transformed) arbitrary observation z and the (transformed) observation z_i and a kernel window

$$w_k(z, z_i) = K\left(\frac{||z - z_i||}{d_k(z)}\right)$$

72

where the kernel width parameter $d_k(z)$ is locally adaptive to the density at z. It is chosen as the distance to the kth nearest neighbour $z_{(k)}$ of z, i.e.,

$$d_k(z) = ||z - z_{(k)}||.$$

The order k of the nearest neighbourhood is considered as a flexible parameter of the algorithm.

In contrast to the use of localizing in smoothing, here a weighting scheme is used, that is based on the distance in the predictor space, instead of the distances in the space of the original covariates. While the spaces are identical in the linear case, the distances differ for the quadratic case. The performance does not seem to depend strongly on the type of weighting. This might be due to the strong relation between the distances in predictor and in covariate space. By partitioning the vectors into $z' = (1, x', \tilde{x}')$, $z_i' = (1, x_i', \tilde{x}_i')$, where $\tilde{x}$ and $\tilde{x}_i$ contain the corresponding quadratic terms, one obtains the simple relation $||z - z_i||^2 = ||x - x_i||^2 + ||\tilde{x} - \tilde{x}_i||^2$, which implies that the distance between z and z_i increases with the distance between x and x_i. The use of distances in the predictor space is preferred because it is easier to handle.

The kernel function K is one component of the localized logistic regression procedure that needs to be chosen. For the further investigations the Gaussian kernel

$$K_G(x) = \exp(-x^2)$$

and the tricube kernel

$$K_T(x) = \begin{cases} (1 - |x|^3)^3 & \text{for} |x| < 1 \\ 0 & \text{otherwise} \end{cases}$$

are considered. Parameter estimation was found to be more stable with the Gaussian kernel because even points far away from x receive a non-zero weight. Again, for localization methods the tricube kernel has the advantage that in estimation only the points in the neighbourhood get included (since all other points receive weight zero). The tricube kernel in effect uses only the k nearest neighbours of z which leads to a distinct computational advantage. In addition better performance is expected with a very local structure.

Parameter estimation is performed by solving the local score equation $s_{x,k}(\beta) = 0$ by iterative Fisher scoring of the form

$$\hat{\beta}_x^{(s+1)} = \hat{\beta}_x^{(s)} + F_{x,k}(\hat{\beta}_x^{(s)})^{-1} s_{x,k}(\hat{\beta}_x^{(s)}) \tag{3.2}$$

where $s_{x,k}(\beta) = \partial l_x / \partial \beta$ is the local score function, which for the logistic model has the simple form

$$s_{x,k}(\beta) = \sum_i w_k(z, z_i) z_i (y_i - \pi_i(\beta))$$

with $\pi_i(\beta)$ denoting the response probability evaluated at β, and $F_{x,k} = E(-\partial^2 l_x / \partial\beta\partial\beta')$ denoting the weighted Fisher matrix

$$F_{x,k} = \sum_i w_k(z, z_i) z_i z_i' \frac{\partial h(\eta_i)}{\partial \eta}$$

with $\eta_i = z_i'\beta$ and $h(x) = \frac{\exp(x)}{1+\exp(x)}$ being the response function of the logistic regression model. The dependence of the parameter estimates on the target value becomes apparent in the notation "$\hat{\beta}_x$". For the asymptotic behavior of local estimates see Fan and Gijbels (1996).

3.1.2 Local reduction of dimensions by selection of predictors

A second step in localizing the logistic regression model is to do local selection of predictors. This is based on the assumption that not all predictors are equally informative with respect to class membership throughout the space spanned by all predictors. Using a global technique for selection would potentially discard predictors which are essential at only one specific point in the space spanned by the predictors, and which are irrelevant elsewhere. There are various procedures that provide measures of relevance of predictors (for an overview see e.g. Liu and Motoda, 1998). For localized logistic regression it is suggested to determine the relevance by the coefficients of the fitted local model. Therefore, the studentized value

$$c_{x,k}(\hat{\beta}_{x,j}) = \frac{|\hat{\beta}_{x,j}|}{\sqrt{\hat{\mathrm{var}}(\hat{\beta}_{x,j})}} \quad , \quad j = 1, \ldots, m,$$

is employed, where $\beta_{x,0}, \beta_{x,1}, \ldots, \beta_{x,m}$ are the elements of β_x. It is based on the variance approximation

$$\hat{\mathrm{cov}}(\hat{\beta}_x) = F_{x,k}(\hat{\beta}_x)^{-1}$$

for the local estimates $\hat{\beta}_x$ at target value x (see Kauermann and Tutz, 2000). This is a localized variant of the Wald statistics for testing the null hypothesis

$\beta_j = 0$ locally. A predictor is retained if the corresponding $c_{x,k}(\hat{\beta}_j)$ exceeds a value c_β. Use of a fixed value $c_\beta = 0.5$ will be investigated, but also a variant of localized logistic regression that treats c_β as the second flexible parameter of the algorithm. After the number of predictors has been reduced the weights $w(z, z_i)$ are recalculated for the subspace spanned by the selected predictors, and the estimation is performed for the reduced β-vector of the final local model. Prediction for target value x is based on the reduced and re-estimated model.

An unanswered question finally is, whether selection should be done by a one-step selection retaining only predictors with standardized coefficients above a certain threshold, or by a stepwise exclusion of predictors with small standardized coefficients and a re-estimation in each step. Since the latter, computationally much more complex, stepwise procedure has not been superior to the one-step selection in initial experiments it was decided in favor of the computationally more attractive one-step procedure. It belongs to the class of supervised techniques, where the response values of the training samples are explicitly used for selection of predictors. In contrast, unsupervised techniques do not require the response values. Bishop (1995) gives an overview of techniques from both categories. Unsupervised techniques that have been adapted to local estimation are locally weighted factor analysis or locally weighted principal component analysis (Schaal et al., 1998), for example.

3.1.3 Computational optimization

There are several problems with parameter estimation for logistic regression that need to be considered. A localized variant is likely to be affected even more by such problems, because often there will be much less data available compared to a global estimation variant. One problem is that parameters tend towards infinity when there is (local) complete or quasi-complete separability, i.e., when the fitted class boundary separates instances from the two classes perfectly for the training data set (Albert and Anderson, 1984). The algorithm introduced here deals with such situations by stopping iterations when $\pi_i(\hat{\beta})$ gets too close to 0 or 1. In this case the estimates from that stage are used, but it is refrained from using the variance estimates and so no selection of predictors is done.

Additional numerical problems arise from (local) collinearities of the predictors. These often cause instability and large estimates of parameters. To avoid such problems a penalization of the parameter estimates is introduced by adding a penalty term to the likelihood (3.1). The resulting penalized weighted log-

75

likelihood is

$$l(\beta) = \sum_i \left(y_i \log \pi(x_i) + (1 - y_i) \log(1 - \pi(x_i)) \right) w_k(z, z_i) - \lambda \beta' I \beta \qquad (3.3)$$

where I is a identity matrix and λ determines the strength of the penalization. By setting λ to 0 the un-penalized likelihood (3.1) is recovered. The proposed penalty term is equivalent to local penalization of β_j^2, which is closely connected to (logistic) ridge regression (Hoerl and Kennard, 1970; Le Cessie and van Houwelingen, 1992). The penalty parameter λ is the third flexible parameter employed in the algorithm.

The modifed expressions for the penalized weighted local score function and Fisher matrix are

$$s_{x,k}(\beta) = \sum_i w_k(z, z_i) z_i (y_i - \pi_i(\beta)) - 2\lambda I \beta$$

and

$$F_{x,k} = \sum_i w_k(z, z_i) z_i z_i' + 2\lambda I.$$

For estimation special attention has to be given to the intercept parameter β_0, which in a localized logistic classification model reflects the local class membership probability. This special parameter cannot be penalized for estimation, because this would shift the estimate towards zero and the corresponding membership probability of 0.5 is not optimal if the local model is fitted for an observation that lies far from the class boundary. Again, without penalization convergence problems arise. Le Cessie and van Houwelingen (1992) use centered predictors (in addition to the standardization that is necessary because of the penalization), and set the intercept to a fixed value. As for localized procedures the weights have to be taken into account, the predictors here are centered and standardized in a weighted way, and the intercept is kept fixed at the value of the transformed local class membership proportion.

It should be noted that usage of penalized estimation alone will already perform some kind of variable selection. While ridge regression initially was introduced for stabilizing estimation by Hoerl and Kennard (1970), later the interest shifted to its variable selection effect. For example, Tibshirani (1996) developed the Lasso algorithm where $|\beta_j|$ is penalized instead of β_j^2. This algorithm results in parameter estimates that are not only smaller, but that are equal to zero for irrelevant predictors. While generalizations of the Lasso such as LARS (Efron et al., 2004) and the elastic net (Zou and Hastie, 2005) have been developed, the generalized linear model framework, including logistic regression, has received less attention and therefore application to the current setting would have been quite costly. Nevertheless, a variant of localized logistic regression will be considered that uses only penalization, and no other kind of variable selection.

3.1.4 Algorithm and parameter selection

For a given measurement x and for fixed parameters of the algorithm k, c_β and λ, the estimation and prediction procedure may be summarized into:

1. Determine $d_k(z)$; calculate weights $w_k(z, z_i) = K(\frac{\|z-z_i\|}{d_k(z)})$.

2. Use iterative Fisher scoring with penalized weighted score function and Fisher matrix (with penalty λ) to determine $\hat{\beta}$.

3. If Fisher scoring converges: Use a subset of predictors where $c_{x,k}(\hat{\beta}_j) > c_\beta$, re-calculate $w_k(z, z_i)$ for that subspace and repeat the Fisher scoring with that subset of predictors.

4. Use the (reduced) model to predict the class for x.

In order to obtain an applicable algorithm that does not suffer from ad hoc choices a fully automatic choice of the parameters of the algorithm based on cross-validation is suggested. The parameters of localized logistic regression are k (index of the neighbour determining the window size), c_β (cutoff for predictor selection) and λ (penalty on parameters). A grid search for the optimal parameters is employed, where each combination is evaluated with leave-one-out cross-validation. The parameters k, c_β, and λ that result in a minimal error rate are subsequently adapted for prediction on new data. The computational cost of leave-one-out cross-validation may seem prohibitive, but as weights and local models have to be calculated for each point separately other cross-validation schemes would not be advantageous. Furthermore, parameter search has to be performed only once at training, and afterwards the parameters found can be used for prediction given that the basic structure of the data remains the same.

3.1.5 Relevance of variables

Users of classifiers are not only interested in the performance of classifiers in terms of misclassification rates. Often they want to know which variables are relevant, and which therefore have to be collected in the future. An advantage of simple parametric classification, like Fisher's linear discriminant analysis or (global) logistic discrimination, is that the relevance of variables can be evaluated by considering the parameter estimates. For nonparametric approaches, or advanced procedures like boosting, the impact of variables becomes much harder to evaluate. For possible ways in boosting see Friedman (2001).

In the case of linear localizing, the approach presented here is explicitly based on variable selection, but in a localized way. The underlying assumption is that different variables are relevant at different points in predictor space. Based on the distribution of predictor values (or their empirical equivalent, the data $x_1, \ldots, x_{n_L}$) one might construct global measures for the relevance of covariates. By considering

$$
I_j(x) = \begin{cases} 1 & \text{if variable } x_j \text{ is selected for prediction of } x \\ 0 & \text{otherwise} \end{cases}
$$

one obtains the simple relevance score

$$
r_j = \frac{1}{n_T} \sum_{i=1}^{n_T} I_j(x_i)
$$

where n_T is the number of points $x \in \{x_1, \ldots, n_T\}$ for which a prediction is wanted. This measure reflects how often variable x_j is considered to be relevant in the observed predictor space. Instead of single variables one might also consider the relevance of combinations or subsets of variables by defining $I_S(x)$ as the indicator function for selection of the subset $S \subset \{x_1, \ldots, x_p\}$.

3.2 Simulation study

In the following the localized logistic regression (LLR) algorithm introduced above will be compared to several other classification procedures by evaluating their performance on test sets which are strictly independent from the training sets. The the following versions of LLR are used:

78

- cLLR: classic linear localized logistic regression (see e.g. Loader, 1999), i.e., no penalisation or selection of predictors. The Gauss kernel is used for stability of estimation. Kernel width is chosen by cross-validation, but very local models often can not be evaluated due to numerical problems.

- pLLR: penalized linear localized logistic regression without selection of predictors, using the tricube kernel

- pLLR Gauss: pLLR using the Gauss kernel.

- pqLLR: a quadratic version of pLLR.

- lLLR: linear localized logistic regression using penalized estimation and selection of predictors.

- lLLR fixed: lLLR with fixed $c_\beta = 0.5$.

- qLLR: a quadratic version of lLLR.

The following procedures will be used for comparison:

- LDA: Linear discriminant analysis.

- NNet: Single-hidden-layer neural networks with the number of hidden units and the size of weight decay being chosen by 10-fold cross-validation. Classification is done by committee voting after training five networks with different starting values (as suggested for example by Venables and Ripley, 1999).

- 1-NN: 1-nearest-neighbourhood classification (see e.g. Fix and Hodges, 1951).

- 10-NN: 10-nearest-neighbourhood classification.

- k-NN: nearest-neighbourhood classification with neighbourhood size chosen by leave-one-out cross-validation.

- Tree: Classification trees (Breiman et al., 1984; Ripley, 1996) with tree size determined by 10-fold cross-validation.

- Bag: Bagging with classification trees (Breiman, 1996).

- RF: Random forests (Breiman, 2001, 2002).

79

These procedures have been chosen in order to be representative of linear, partition-based and model-free methods. So it will be instructive to investigate in what situations LLR performance is similar to any of the other methods. The variants of nearest-neighbourhood classification with fixed size of the neighbourhood will give an indication whether a classification problem is rather local (when 1-NN performs better) or rather global (when 10-NN performs better). The implementations used are those from the statistical environment R (R Development Core Team, 2005) and the standard settings for all procedures were used. For all variants of localized logistic regression self-developed implementations were used.

Values of parameters k, c_β, and λ that lead to optimal cross-validation scores are obtained by a search on a $10 \times 6 \times 6$ grid. Here, optimization of these parameters is done in a global way, i.e., cross-validation is used to find one set of parameters that will be used for all predictions derived from that one specific set of training data. This global optimization approach implies that the same amount of localization, predictor selection and penalization is necessary everywhere in predictor space. To evaluate this assumption, local optimization based on a weighted version of AIC (see e.g. Loader, 1999) was investigated. As this did not result in a consistent improvement in initial experiments, and given the additional computational burden that results from local optimization at the time of prediction, global optimization is retained.

3.2.1 Example data and results

A variety of types of data will be used, following Friedman (1994) and Hastie and Tibshirani (1996). The examples vary with respect to importance of variables, number of noise variables, distribution of variables per class, and the shape of the class regions. Presentation distinguishes between normally distributed classes, examples with non-overlapping class regions, and examples with fractioned class regions, i.e., class distributions that are disjoint. For each example 50 repetitions of data generation, model fitting, and prediction have been done.

The size of the training data was set to $n_L = 200$ and the number of observations in the test data was set to $n_T = 1000$ for the majority of the examples. Both are equally divided between the two classes used in each example. Performance of the procedures is evaluated by prediction error on the test data. For illustration often the relative error rate will be given, meaning that for each repetition the error rate of each procedure is divided by the smallest error rate which for this repetition is achieved by any of the classification methods under comparison. So

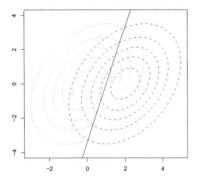

Figure 3.2: Class densities for the two classes from example HT1 (isodensity contours: broken curves and dotted curves) and Bayes decision boundary (solid line).

for example a procedure that is the best in each repetition would always have the relative error rate one. This type of illustration is used by Friedman (2001) for example, and makes it easier to judge relative performance.

Classes with covariates from multivariate normal distributions

To start a simple example (HT1) is used, which is equivalent to example 1 of Hastie and Tibshirani (1996), in which two covariates are drawn from a normal distribution with variance given by $\text{var}(x_1) = 1$, $\text{var}(x_2) = 2$, and correlation 0.75. The mean of the two classes is separated by two units on the first dimension. Figure 3.2 gives the density for the two classes (broken curves and dotted curves) and the Bayes decision boundary (solid curve). These were obtained by kernel density estimation for a large number of observations. The Bayes decision boundary, i.e., the boundary which gives the minimum expected prediction error based on the class densities, is linear for this example.

The left panel of Figure 3.3 shows the mean relative error rates for the aforementioned example for localized logistic regression procedures without selection of predictors, localized logistic regression with selection of predictors, and for the other procedures used for comparison. Linear discriminant analysis (LDA) appears to have the best performance. This does not come as a surprise be-

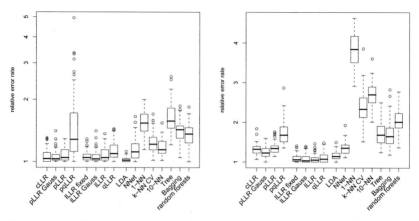

Figure 3.3: Relative error rates for two classes with a 2-dimensional normal distribution with non-zero covariance. Left panel (note the log-scale): Two covariates (HT1). Right panel: Two covariates with additional 14 noise covariates drawn from a standard normal distribution (HT2).

cause the decision boundary is a straight line and can be matched very well by LDA. All variants of localized logistic regression (LLR), with the exception of quadratic localized logistic regression without selection of predictors (pqLLR), have performance relatively close to LDA. The mean of the chosen parameters for linear localized logistic regression with selection of predictors using the tricube kernel (lLLR) is $k = 0.8 \cdot n_L$, $c_\beta = 0.828$, and $\lambda = 0.554$. This indicates that in most of the repetitions the optimization procedure chooses parameters that result in no localization, and so LLR becomes a global logistic discrimination procedure which resembles LDA. All LLR procedures that use the Gauss kernel perform slightly better compared to their tricube kernel counterparts. For quadratic localized logistic regression this example demonstrates the usefulness of the predictor selection step. Omission of this component results in a drastic degradation of performance (compare pqLLR to qLLR). When selection of predictors is added, the local fitting of a quadratic model becomes rather stable. Its performance is slightly worse than that of linear localized logistic regression, but still it outperforms most of the other procedures.

To test the variable selection component of localized logistic regression, the two covariates providing information on class membership are augmented by 14 noise variables taken from a standard normal distribution (HT2), as in example 2 of Hastie and Tibshirani (1996). As can be seen in the right panel of Figure 3.3 the performance of LLR procedures that do not employ selection of predictors is

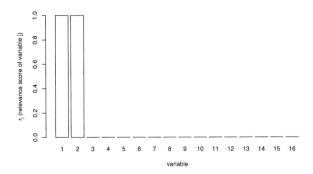

Figure 3.4: lLLR relevance scores of covariates for an example where the first two variables carry information and the other 14 are noise variables (lLLR parameters: $k = n_L$, $c_\beta = 1.6$ and $\lambda = 0.42$).

distinctly worse than that of LLR with selection of predictors. Similar to example HT1 this difference is especially large for quadratic localized logistic regression. The performance of quadratic LLR with selection of predictors is similar to the linear version. This indicates that even in the presence of a considerable amount of noise the superfluous complexity of the quadratic version does not result in overfitting, but is reduced to a level that is appropriate for the underlying structure. The means of the chosen parameters for lLLR $k = 0.972 \cdot n_L$, $c_\beta = 0.680$ and $\lambda = 0.378$ indicate that still a rather global model is used.

Figure 3.4 shows the covariate relevance scores obtained from lLLR for one repetition from example HT2. It can be seen that variables that carry information on class membership have been distinctly identified. This result taken together with the good performance of lLLR indicates that selection of predictors succeeds here. Another indicator for the usefulness of local variable selection can be seen in the worsening performance of the nearest-neighbourhood algorithms due to the addition of noise variables. They share the notion of localization with the LLR procedures, but lack the possibility of predictor selection.

In the next two examples (F1 and F2), equivalent to examples 1 and 2 of Friedman (1994), the amount of information provided by the variables is systematically varied. For both classes the covariates are drawn from a 10-dimensional normal distribution. For the class with $y_i = 0$ the data are generated from standard normal distributions. For the other class the mean and the variance depend on

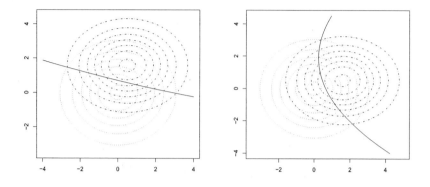

Figure 3.5: Class densities for the two classes and covariates 1 and 10 from example F1 (left panel) and F2 (right panel) (isodensity contours: broken curves and dotted curves) and Bayes decision boundary (solid curves).

the variable index. Covariance is set to zero in both examples. In the first example (F1) the covariates with the higher index j are intended to be more relevant. So for $y_i = 1$ data are generated from a normal distribution $x_i \sim N(m, C)$ where

$$\{m_j = \sqrt{j}/2\}_1^p, \ C = \mathrm{diag}\{1/\sqrt{j}\}_1^p.$$

The left panel of Figure 3.5 shows the class densities and the Bayes decision boundary for covariates 1 and 10. A linear decision boundary seems to be adequate for this example. This is also reflected by the relative error rates shown in the left panel of Figure 3.6. The good performance of several procedures that use a linear combination of predictors (e.g. pLLR, lLLR and LDA) indicates that indeed the Bayes decisions boundary can be approximated very well by hyperplanes. Nevertheless, the performance of cLLR and pqLLR is indicative of difficulties concerning the large number of covariates. Taking into account that for example Loader (1999) used classic localized logistic regression for a maximum of four predictors, this does not come as a surprise. At least for pqLLR performance can be improved by adding selection of predictors (resulting in qLLR). On the other hand, the performance for linear LLR seems to decrease when selection of predictors is used. The better performance of LLR with a fixed level c_β indicates that flexible selection of c_β might be problematic in this example.

84

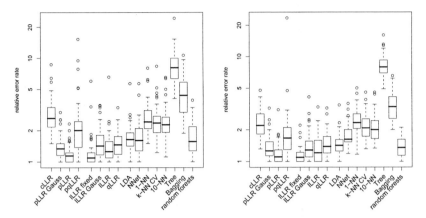

Figure 3.6: Relative error rates (note the log-scale) for data with covariates from 10-dimensional normal distributions. Left panel: Variables with higher index carry more information on class membership (F1). Right panel: Variables with low index have more information due to the mean, variables with higher index have more information due to the variance (F2).

In the second example (F2) the mean structure of the second class is changed to

$$\{m_j = \sqrt{p - j + 1/2}\}_1^p$$

and so the variables with lower index contain more relevant information on class membership due to the mean, and the variables with higher index due to the variance. The resulting effect is that the Bayes decision boundary becomes more complex (right panel of Figure 3.5). Nevertheless, as can be seen from the the relative error rates in the right panel of Figure 3.6, procedures that employ linear combinations of predictors have the best performance (again with the exception of cLLR and pqLLR). While quadratic LLR in combination with predictor selection performs very stable, it is noteworthy that there is no additional benefit of the quadratic terms for fitting the curved class boundary. As there are 20 predictors (resulting from 10 covariates) compared to only 200 observations, this might be due to the large number of parameters to be estimated. On the other hand the means of the chosen parameters for lLLR ($k = 0.942 \cdot n_L$, $c_\beta = 0.640$, and $\lambda = 0.756$ for F1 and $k = 0.864 \cdot n_L$, $c_\beta = 0.628$, and $\lambda = 0.756$ for F2) indicate that for linear LLR a global linear class boundary is no longer sufficient.

Classes with non-overlapping connected class regions

In the examples presented in this section the variables that carry information on class membership define non-overlapping and connected class regions. In some examples they are augmented by noise variables. In the first example (F5), which is equivalent to example 5 of Friedman (1994), the class boundary is defined by a linear combination of the covariates. It is constructed in a way so that all input variables have equal local relevance everywhere in the space spanned by the covariates. However, there is a single direction in that space that contains all the discriminating information. There are $p = 10$ covariates with class membership rule

$$\sum_{j=1}^{10} x_{ij} \leq 9.8 \Rightarrow y_i = 0, \text{ otherwise } y_i = 1.$$

The left panel of Figure 3.7 shows the relative error rates for 50 repetitions with $n_L = 200$ and $n_T = 1000$. The procedures that utilize a linear combination of predictors clearly have the best performance (with all LLR procedures among them). The simple data structure of the example is emphasized by the fact that classic localized regression, i.e., LLR without any refinements, performs best. Nevertheless the flexibility of the other variants is controlled adequately, indicated for example by the means of the chosen parameters for lLLR $k = 0.996 \cdot n_L$, $c_\beta = 0.848$, and $\lambda = 0.279$. Again quadratic LLR benefits the most from selection of predictors.

The results are quite different for example F4 (as in example 4 of Friedman (1994)) in which a quadratic combination of covariates is used to define the class boundary: The right panel of Figure 3.7 shows the results for the class membership rule

$$\sum_{j=1}^{10} x_{ij}^2 \leq 9.8 \Rightarrow y_i = 0, \text{ otherwise } y_i = 1$$

and a training sample size of $n_L = 500$ and $n_T = 1000$. The performance in the case of classic LLR, lLLR with fixed c_β, and of the linear procedures LDA, and of neural networks degrades. The best performance can be seen for lLLR and qLLR, and for their variants without selection of predictors. Only performance of neural networks, bagging, and random forests is comparable to that of local linear procedures. Of course quadratic LLR is the distinct winner in this example, because by utilizing quadratic components the quadratic class boundary can be approximated very well even by usage of a global model (as indicated by the means of the chosen parameters for quadratic localized logistic regression $k = n_L$, $c_\beta =$

86

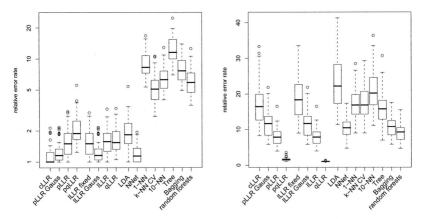

Figure 3.7: Relative error rates for classes with non-overlapping connected class regions defined by a linear (example F5; left panel; note the log-scale) or quadratic (example F4; right panel) combination of covariates.

0.952, and $\lambda = 0.599$). It should be noted that nevertheless *linear* localization works very well even in this case. The means of the chosen parameters for lLLR $k = 0.594 \cdot n_L$, $c_\beta = 0$, and $\lambda = 0.833$ indicate that rather local models are chosen to approximate the quadratic structure. The small cutoff for variable selection indicates that the standardized parameter estimates are very small, but still the procedure detects the relevance and decides to include all variables into the model. In contrast lLLR with fixed cutoff for selection of predictors seems to exclude relevant predictors, and therefore its performance is much worse, even compared to LLR without selection of predictors. So in this example it becomes very important to treat the cutoff c_β as a flexible parameter.

A variant of the considered data structure is characterized by varying relevance of predictors. In example F3, as in example 3 of Friedman (1994), a weighted contribution of the variables

$$\sum_{j=1}^{10} x_{ij}^2/j \le 2.5 \Rightarrow y_i = 0, \text{ otherwise } y_i = 1$$

with training sample size of $n_L = 200$ and $n_T = 1000$ is used. The left panel of Figure 3.8 shows the corresponding class densities and Bayes decision boundary for covariates 1 and 10. As can be seen from the relative error rates in the right panel of Figure 3.8 the performance of pLLR and lLLR degrades to the level of

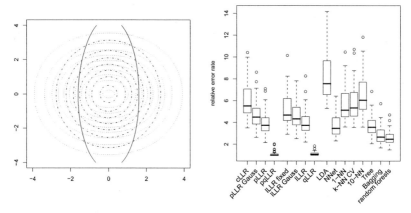

Figure 3.8: Left panel: Class densities for the two classes and covariates 1 and 10 from example F3 (isodensity contours: broken curves and dotted curves) and Bayes decision boundary (solid curves). Right panel: Relative error rates for various classification procedures.

neural networks. This indicates that the class boundary might be too complicated now to be approximated very well by a local linear model (taking into account also the reduced training sample size compared to example F4). The means of the chosen parameters for lLLR $k = 0.498 \cdot n_L$, $c_\beta = 0.240$, and $\lambda = 0.718$ support this reasoning as still an amount of localization similar to F4 is used, but the criterion for variable selection is more strict. As all variables are relevant, such a strict criterion for exclusion of predictors is not appropriate. In contrast quadratic LLR again performs very well due to the quadratic approximation of the class boundary.

In the next three examples (HT5), based on example 5 of Hastie and Tibshirani (1996), there are four covariates drawn from standard normal distributions. Class membership is assigned by the rule

$$\sqrt{\sum_{j=1}^{4} x_{ij}^2} \leq 3 \Rightarrow y_i = 0, \text{ otherwise } y_i = 1.$$

These variables are augmented with a varying number of noise variables drawn from a standard normal distribution. The left panel of Figure 3.9 shows the relative error rates for an example with no noise variables. The performance of lLLR is very satisfying here, employing rather local models and a weak criterion for variable selection ($k = 0.476 \cdot n_L$, $c_\beta = 0.024$, and $\lambda = 0.756$). The low cutoff for

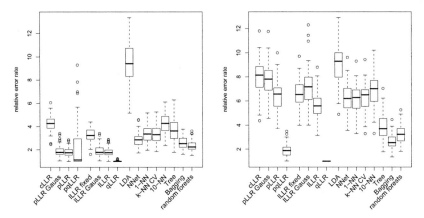

Figure 3.9: Relative error rates for examples where the class region of one class is a four-dimensional sphere (HT5) without any noise variables (left panel) or with sixteen noise variables in addition (right panel).

selection of predictors that seems to be required is likely to be the reason for the poor performance of LLR with fixed cutoff. While in example HT2 the fixed value $c_\beta = 0.5$ was adequate, in this example it is too strict, indicating that a flexible selection is needed. For quadratic LLR a distinct benefit of predictor selection can be seen. Without selection it shows large variability, with predictor selection it performs best of all procedures. When adding six standard normal noise variables the performance of linear LLR (with means of the chosen parameters $k = 0.546 \cdot n_L$, $c_\beta = 0.172$, and $\lambda = 0.746$) decreases relative to procedures such as bagging and random forests (shown later in Tables 3.1 and 3.2). As seen from the right panel of Figure 3.9, with 16 noise variables the performance decreases even more. The means of the chosen parameters $k = 0.462 \cdot n_L$, $c_\beta = 0.248$, and $\lambda = 0.753$ indicate that an increasing amount of noise enforces a more strict criterion for variable selection, and so variables with corresponding small standardized parameter estimates are falsely excluded. Nevertheless, comparing the performance of pLLR and lLLR it can be seen that there is a benefit of performing selection of predictors. The performance of qLLR relative to other procedures does not seem to be impaired by the inclusion of noise variables. This indicates that selection of predictors works reliable if the local models have the appropriate structure.

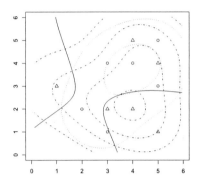

Figure 3.10: Class densities for the two classes from example HT3 (isodensity contours: broken curves and dotted curves) and Bayes decision boundary (solid curves) for one set of subclass centres (circles and triangles).

Fractioned class structure

The examples presented up to now used classes in which either observations were drawn from one distribution per class or in which one connected class region was used. In the following, examples will be considered with extremely fractioned class regions, i.e., class distributions that are disjoint, denoted by HT3 and HT4 similar to examples 3 and 4 of Hastie and Tibshirani (1996).

In the first example (HT3) the distribution of each of the two classes is defined as a mixture of six spherical bivariate normal subclasses. The standard deviation of each subclass is 0.25. The means of the 12 subclasses are chosen for each repetition at random (without replacement) from the integers $[1, 2, \ldots, 5] \times [1, 2, \ldots, 5]$. There are 20 observations drawn from the distribution of each subclass and so there are 140 observations per class with a total of $n_L = 240$ observations ($n_T = 960$). Figure 3.10 gives the estimated class densities, estimated Bayes decision boundaries, and subclass means (indicated by circles and triangles) for one set of data generated according to this rule. Procedures like LDA that allow only for a single (linear) class boundary can hardly adequately capture this structure. This indeed happens as can be seen from the left panel of Figure 3.11. All other procedures (except pqLLR and lLLR fixed) perform very similar. The best performance is found for lLLR and 10-nearest neighbourhood. The selec-

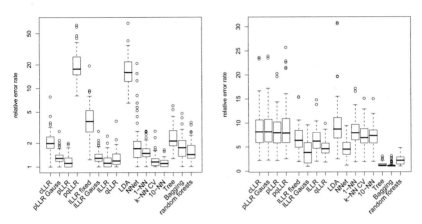

Figure 3.11: Relative error rates for an extremely fractioned class structure. Left panel (note the log-scale): without noise variables (HT3). Right panel: with eight noise variables (in addition to two variables carrying information on class membership) (HT4).

tion of the localization parameter by cross-validation for lLLR (the means of the selected parameters being $k = 0.180 \cdot n_L$, $c_\beta = 0.108$, and $\lambda = 0.654$) results in very local models for this example. This demonstrates that LLR can become a nearest neighbourhood method in the limiting case. Classic LLR which cannot fit such local models due to issues with numeric stability can be seen to have a performance that is much worse. Although there are no noise variables in this example, quadratic LLR is seen to perform well only with selection of predictors.

In the next example (HT4) the two variables carrying information on class membership are augmented with eight noise covariates having standard normal distribution. As can be seen in the right panel of Figure 3.11 the performance of LLR procedures degrades compared to the partition-based classification tree, bagging and random forest procedures. The latter procedures seem to perform very well in separating informative variables from noise variables. While selection of predictors results in some performance improvement for lLLR, by looking at which covariates got selected (Figure 3.12) it is clear that the selection did not work very well for the local models. The lack of detection of adequacy of local models, and the failure to include variables that carry information is also reflected by the means of the chosen parameters when compared to example HT3: $k = 0.544 \cdot n_L$, $c_\beta = 0.464$, and $\lambda = 0.538$. These problems may arise due to the relatively low number of data points for each subclass compared to the number of noise variables. The largest performance benefit of predictor selection can be seen for

91

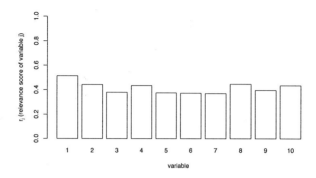

Figure 3.12: lLLR relevance scores of variables for data with extremely fractioned class regions defined by two variables augmented by eight noise variables. Only the first two variables carry information on class membership. (lLLR parameters: $k = 0.2 \cdot n_L$, $c_\beta = 0.4$ and $\lambda = 0.58$)

LLR with the Gauss kernel. While showing a large variability, this is the only procedure that comes close to the partition-based procedures.

3.2.2 Summary of simulation results

Table 3.1 gives the mean absolute error rates for all of the examples and all LLR procedures evaluated. The results for the best two procedures (or more, in the case of ties) are printed in boldface for each example. It can be seen that in all but the simplest examples (HT1 and F5) classical localized logistic regression is outperformed by the more refined variants. The improvement is largest with complicated data structure that requires very local models (examples F3 and F4), which cannot be estimated reliably without penalization, or in case there are many noise variables (examples HT5 and HT4). Deciding whether the Gauss kernel or the tricube kernel should be used is not obvious, because their performance depends on the specific examples. There are three examples where the tricube kernel performs distinctly better (F4, F3 and HT5 with 16 noise variables), and only one where the Gauss kernel clearly dominates (HT4). Taking further into consideration the computational advantage the tricube kernel should be preferred.

Benefits of selecting predictors for linear localized logistic regression cannot be reported for all examples. There are some examples, especially with simple data

Table 3.1: Mean error rates for different simulated data examples and various versions of localized logistic regression (LLR) (cLLR: classic LLR; pLLR: LLR with penalized estimation, tricube kernel and no selection of predictors; pLLR G: pLLR with Gauss kernel; pqLLR: quadratic version of pLLR; lLLR: LLR with penalized estimation, tricube kernel and selection of predictors; lLLR f: lLLR with fixed cutoff $c_\beta = 0.5$; lLLR G: lLLR with Gauss kernel; qLLR: quadratic version of lLLR).

	cLLR	pLLR G	pLLR	pqLLR	lLLR f	lLLR G	lLLR	qLLR
multivariate normal								
2-dim (HT1)	**0.067**	**0.067**	0.068	0.102	0.068	**0.067**	0.069	0.072
2-dim with noise (HT2)	0.088	0.083	0.090	0.115	0.073	**0.072**	**0.072**	0.074
10-dim, var. inf. (F1)	0.031	0.016	**0.013**	0.029	**0.014**	0.017	0.016	0.017
10-dim, diff. inf. (F2)	0.035	0.021	**0.019**	0.032	**0.018**	0.022	0.021	0.023
no overlap, connected								
linear combination (F5)	**0.033**	0.035	0.045	0.058	0.044	**0.035**	0.045	0.046
quadratic comb. (F4)	0.343	0.231	0.156	**0.034**	0.371	0.231	0.156	**0.021**
weighted quad. (F3)	0.372	0.300	0.245	**0.074**	0.325	0.297	0.250	**0.074**
quad., no noise (HT5)	0.228	0.101	**0.095**	0.130	0.174	0.102	0.097	**0.057**
quad., some noise (HT5)	0.325	0.284	0.218	**0.072**	0.272	0.256	0.219	**0.056**
quad., more noise (HT5)	0.445	0.433	0.358	**0.107**	0.368	0.405	0.312	**0.058**
fractioned class structure								
without noise (HT3)	0.042	0.027	**0.024**	0.383	0.085	0.027	**0.024**	0.026
with noise (HT4)	0.313	0.311	0.296	0.337	0.253	**0.165**	0.246	**0.194**

structure, for which no performance improvement occurs by using the additional selection step, and for some examples performance even becomes worse. On the other hand, there is no large performance penalty in these examples either, indicating that the selection of the cutoff parameter c_β works sufficiently well, and situations where all predictors should be retained can be detected. For the examples with noise variables (HT2, HT4 and HT5 with 16 noise variables) a distinct performance improvement becomes manifest. Thus, in situations in which one expects only some of the covariates to carry information with respect to class membership, it is save and advised to perform the predictor selection step for linear localized logistic regression.

For quadratic localized logistic regression the benefit of selecting predictors is obvious: Omitting the selection step results in severe overfitting for several examples even if there are no noise variables (HT1, HT3, HT5 with no noise variables), and makes use of quadratic localized logistic regression problematic for examples with noise variables (HT2, HT4, and HT5 with noise variables). So if the additional flexibility of quadratic local models is desired it has to be controlled by selection of predictors.

Table 3.2 gives the mean error rates for linear and quadratic localized logistic regression with selection of predictors, and lists the other procedures used for comparison (excluding 1-NN and 10-NN). The error rates of the two best proce-

Table 3.2: Mean error rates for different simulated data examples and classification procedures (lLLR: linear localized logistic regression; qLLR: quadratic LLR; LDA: linear discriminant analysis; NNet: neural networks with committee voting; k-NN: nearest-neighbourhood classification with neighbourhood size chosen by cross-validation; Tree: cross validated classification trees; Bag: bagging with trees; RF: random forests).

	lLLR	qLLR	LDA	NNet	k-NN	Tree	Bag	RF
multivariate normal								
2-dim (HT1)	**0.069**	0.072	**0.065**	0.072	0.078	0.104	0.090	0.086
2-dim with noise (HT2)	**0.072**	0.074	0.077	0.090	0.157	0.115	0.114	0.135
10-dim, var. inf. (F1)	**0.016**	0.017	0.019	0.019	0.027	0.089	0.050	0.019
10-dim, diff. inf. (F2)	**0.021**	0.023	**0.022**	0.027	0.033	0.124	0.052	0.021
no overlap, connected								
linear combination (F5)	**0.045**	0.046	0.053	**0.034**	0.142	0.329	0.214	0.166
quadratic comb. (F4)	0.156	**0.021**	0.453	0.206	0.344	0.309	0.213	0.179
weighted quad. (F3)	0.250	**0.074**	0.507	0.230	0.358	0.229	0.180	**0.164**
quad., no noise (HT5)	0.097	**0.057**	0.496	0.152	0.178	0.196	0.139	0.126
quad., some noise (HT5)	0.219	**0.056**	0.499	0.239	0.286	0.210	0.146	**0.145**
quad., more noise (HT5)	0.312	**0.058**	0.505	0.345	0.349	0.218	**0.150**	0.180
fractioned class structure								
without noise (HT3)	**0.024**	0.026	0.330	0.048	**0.024**	0.048	0.035	0.031
with noise (HT4)	0.246	0.194	0.342	0.180	0.262	**0.059**	**0.053**	0.096

dures in each example are printed in boldface.

It can be seen that for different situations different classification methods turn out to be the best choice, but some procedures react more flexible to varying data structures. Given that it cannot be expected that one method is superior in all data situations, localized logistic regression (LLR) performs reliable for a variety of different data structures. This will be explained in the following points which summarize over the previous observations:

1. LLR shows better performance than LDA and nearest neighbourhood approaches in almost all examples (one exception for LDA, one exception for 10-NN). In the examples where LLR and LDA have similar performance (HT1 and HT2) the optimal LLR localization parameter with respect to the cross-validation score is found to be $k = n_L$. This indicates that local models are not necessary for these examples, and LLR becomes a global procedure. For the few examples in which nearest neighbourhood methods perform well (e.g. HT3) the performance of LLR is similar to k-NN. LLR parameter selection here is found to favour very local models, and so LLR becomes a nearest neighbourhood method.

2. Surprisingly, the comparison to neural networks is often in favour of LLR, despite the fact that in contrast to LLR neural networks can model interac-

tions of covariates directly. Neural networks as used here perform distinctly better only in two examples. The same holds true for simple trees which offer no serious alternative.

3. In all the examined examples in which observations are drawn from a multivariate normal distribution per class, with and without noise, the performance of LLR is one of the best of all methods. Trees perform very badly, only random forests come close to LLR.

4. For the investigated data structures with non-overlapping class regions the only competitors to localizing techniques are advanced tree methodologies, such as bagging and random forests. For F3 and HT5 (some noise) random forests outperform linear localizing procedures. When quadratic terms are included into localizing, LLR dominates distinctly which in these cases is due to the underlying quadratic structure.

5. For data with fractioned class structure with noise variables, tree-based approaches perform particularly well if noise variables are included. Although LLR performs better without noise variables, it is outperformed if much noise is present. Advanced tree methodology, such as bagging and random forests, clearly perform best in this case.

6. The comparison between linear and quadratic LLR shows that the quadratic version performs only slightly worse in examples in which linear local models work well. This indicates that penalization and predictor selection succeed in preventing overfitting when quadratic local models have superfluous complexity. On the other hand, there are several examples (e.g. F3 and HT5) in which linear LLR is clearly outperformed by quadratic LLR and so the latter should be preferred.

One might argue that these examples may be artificially constructed in order to favor quadratic models, but it can be seen that linear LLR performs well compared to other procedures even with quadratic structure in some examples (e.g. F4). So the examples illustrate for which types of data structure linear local models are sufficient, and for which they are not.

3.3   Real data

In the previous section simulated examples have been used to investigate how different types of structure affect the performance of localized logistic regression (LLR). As simulated data are by definition always artificial, in this section real data sets will be used to investigate real world performance.

The Australian credit data from the Statlog project (Michie et al., 1994), and the breast cancer and the sonar data from the UCI machine learning repository (Blake and Merz, 1998) will be used. One reason for this selection of data sets is that they have been employed in a recent work on boosting methods (Bühlmann and Yu, 2003), and therefore information on error rates is available for a class of procedures that is considered to perform very well.

For the Australian credit data the aim is to devise a rule for assessing applications for credit cards. The data set has 14 covariates and 690 observations. Due to confidentiality neither the meaning of the covariates nor the exact meaning of the two classes is available. For the use with LLR, LDA, neural networks, and the nearest neighbourhood methods, some variables needed to be transformed from categorical to binary dummy variables, with categories of relative frequency below five percent being discarded. The sonar data contains 208 sonar patterns from either "mines" or "rocks" at various angles and under various conditions. Each pattern comprises a set of 60 numbers in the range 0.0 to 1.0. The aim is to classify an object as "mine" or "rock" given a pattern. The breast cancer data has nine predictors and 699 observations. The classes are "benign" and "malignant", and the covariates contain various cell characteristics.

Each data set has been split 50 times randomly into a 90% training and 10% test set, and all procedures used in the simulation study have been applied. Linear instead of quadratic LLR is used, because it is much faster and shows sufficently good performance. Table 3.3 denotes the error rates for the three data sets and all procedures used in the simulation study. In addition the error rates for several boosting procedures, as given in Bühlmann and Yu (2003), are shown. For two data sets results for boosting with splines in addition to boosting with tree stumps are available.

For the Australian credit data nearest neighbourhood classification rules yield very bad performance, while the rest of the procedures are well comparable. For the breast cancer example lLLR, 10-nearest neighbourhood classification, and random forests distinctly outperform the rest. It can be seen that lLLR performs well for all three data sets. Special attention should be given to the superior performance of lLLR for the sonar data. The relatively good performance of 1-nearest neighbourhood compared to 10-nearest neighbourhood classification hints at a very local data structure. The good performance of neural networks and that of boosting with splines compared to boosting with tree stumps indicates that there is some kind of linear structure. LLR is able to model local as well as linear structures. This combination might explain for its superior performance.

Table 3.3: Error rates for real data and various classification procedures. The numbers are mean error rates for 50 random splits into a 90% training and 10% test set. (lLLR: linear localized logistic regression; LDA: linear discriminant analysis; NNet: neural networks with committee voting; 1-NN/10-NN: 1- and 10-nearest-neighbourhood classification; k-NN: nearest-neighbourhood classification with neighbourhood size chosen by cross-validation; Tree: cross-validated classification trees; Bag: bagging with trees; RF: random forests; L2Boost, L2WCBoost and LogitBoost: various boosting algorithms with tree stump and spline base learners).

	Australian credit	breast cancer	sonar
lLLR	0.126	0.029	0.078
LDA	0.146	0.037	0.273
NNet	0.140	0.035	0.165
1-NN	0.325	0.039	0.181
10-NN	0.313	0.029	0.336
k-NN	0.326	0.029	0.206
Tree	0.153	0.054	0.271
Bag	0.138	0.039	0.212
RF	0.125	0.028	0.164
L2Boost*	0.123	0.037	0.228
with spline*		0.036	0.178
L2WCBoost*	0.123	0.040	0.190
with spline*		0.043	0.168
LogitBoost*	0.131	0.039	0.158
with spline*		0.038	0.148

* from Bühlmann and Yu (2003)

Although averaging across various splits is preferable for the sonar data in addition a specific 50% split is used because it is a reference suggested by Gorman and Sejnowski (1988) and has been used by Hastie and Tibshirani (1996). Due to the different setup the error rates obtained cannot be compared directly to those in Table 3.3 but nevertheless are very similar (with the exception of 1-nearest neighbourhood): 0.106 (lLLR), 0.240 (LDA), 0.115 (neural networks), 0.087 (1-nearest neighbourhood), 0.288 (10-nearest neighbourhood), 0.269 (trees), 0.192 (bagging) and 0.173 (random forests). Thus also for the fixed splitting LLR performs very well. Hastie and Tibshirani (1996) obtained for their discriminant adaptive nearest neighbour classifier (DANN) the test error rate 0.048 which is better than the LLR procedure. When a finer grid is used for parameter selection the LLR error rate reduces to 0.010. This may be interpreted as an artifact, but it also shows that with some tuning of the parameter selection procedure the LLR results are well comparable to the results for DANN and other procedures given by Hastie and Tibshirani (1996), which indicates good local adaptivity.

3.4 Discussion

A localized discrimination procedure has been proposed which in combination with local selection of predictors shows promising results. It works well across a variety of simulated data structures. The proposed local selection of predictors has been shown to be crucial for quadratic local models. For linear local models it results in performance improvements when there are many noise variables, and behaves reliably otherwise. For real data sets, the localizing methodology shows the best performance for two of the considered data sets. This reveals its potential in statistical applications. Although a method cannot be expected to be best for all potential data structures, the performance of localized discrimination is surprisingly good over a wide range of data structures. While it outperforms advanced tree-based methodology like bagging and random forests for simple structures, the latter dominate linear localized logistic regression for quadratically separated classes with many noise variables.

It is especially noteworthy that localized logistic regression works also well in cases in which it has superfluous flexibility. For example no serious loss in performance has to be expected, when LDA would be sufficient and nevertheless localized logistic regression is used. Also in the real data example with categorical predictors localized logistic regression performed well despite the fact that binary predictors pose a problem to distance and weight calculation. Apparently there seem to be few examples where one would be ill-advised to use localized logistic regression, and this recommends the method as supplement the statistical toolbox for classification. Nevertheless, really strong performance benefits can only be gained in situations with several metric predictors and with possibly complicated interactions. The automatic modelling of interactions by use of local models can be expected to outperform procedures, in which the number and degree of interactions has to be specified explicitly in advance.

Bibliography

Akaike, H. (1973). Information theory and the maximum likelihood principle. In Petrov, B. N. and Csaki, F., editors, *2nd International Symposium on Information Theory*, Budapest. Akademiai Kiado.

Albert, A. and Anderson, J. A. (1984). On the existence of maximum likelihood estimates in logistic regression models. *Biometrika*, 71(1):1–10.

Arjas, E. and Haara, P. (1987). A logistic regression model for hazard: asymptotic results. *Scandinavian Journal of Statistics*, 14:1–18.

Bellman, R. E. (1961). *Adaptive Control Processes: A Guided Tour*. Princeton University Press, Princeton, NJ.

Binder, H. and Tutz, G. (2004). Localized logistic classification with variable selection. In Antoch, J., editor, *COMPSTAT 2004*, pages 705–712, Heidelberg. Physica-Verlag.

Bishop, C. M. (1995). *Neural Networks for Pattern Recognition*. Clarendon Press, Oxford.

Blake, C. and Merz, C. (1998). UCI repository of machine learning databases.

Bliss, C. I. (1935). The calculation of the dosage-mortalitiy curve. *Annals of Applied Biology*, 22:134–167.

Breiman, L. (1996). Bagging predictors. *Machine Learning*, 24(2):123–140.

Breiman, L. (2001). Random forests. *Machine Learning*, 45(1):5–32.

Breiman, L. (2002). Manual on setting up, using, and understanding random forests v3.1.

Breiman, L., Friedman, J. H., Olshen, R. A., and Stone, C. J. (1984). *Classification and Regression Trees*. Wadsworth.

Bühlmann, P. and Yu, B. (2003). Boosting with the L2 loss: Regression and classification. *Journal of the American Statistical Association*, 98:324–339.

Buja, A., Hastie, T., and Tibshirani, R. (1989). Linear smoothers and additive models. *The Annals of Statistics*, 17(2):453–555.

Chambers, J. M. and Hastie, T. J. (1992). *Statistical Models in S*. Wadsworth, Pacific Grove, California.

Cox, D. R. (1972). Regression models and life tables (with discussion). *Journal of the Royal Statistical Society B*, 34:187–220.

de Boor, C. (1978). *A Practical Guide to Splines*. Springer, New York.

Duda, R. O., Hart, P. E., and Stork, D. G. (2001). *Pattern Classification*. Wiley, 2nd edition.

Efron, B. (1986). How biased is the apparent error rate of a prediction rule? *Journal of the American Statistical Association*, 81(394):461–470.

Efron, B. (1988). Logistic regression, survival analysis, and the kaplan-meier curve. *Jounal of the American Statistical Association*, 83(402):414–425.

Efron, B., Hastie, T., Johnstone, I., and Tibshirani, R. (2004). Least angle regression. *The Annals of Statistics*, 32(2):407–499.

Eilers, P. H. C. and Marx, B. D. (1996). Flexible smoothing with B-splines and penalties. *Statistical Science*, 11(2):89–121.

Eilers, P. H. C. and Marx, B. D. (2002). Generalized linear additive smooth structures. *Journal of Computational and Graphical Statistics*, 11(4):758–783.

Fahrmeir, L. (1994). Dynamic modelling and penalized likelihood estimation for discrete time survival data. *Biometrika*, 81(2):317–330.

Fahrmeir, L. and Lang, S. (2001). Bayesian inference for generalized additive mixed models based on Markov random field priors. *Journal of the Royal Statistical Society: Series C (Applied Statistics)*, 50(2):201–220.

Fahrmeir, L. and Tutz, G. (2001). *Multivariate Statistical Modelling Based on Generalized Linear Models*. Springer, New York, 2nd edition.

Fahrmeir, L. and Wagenpfeil, S. (1996). Smoothing hazard functions and time-varying effects in discrete duration and competing risks models. *Journal of the American Statistical Association*, 91(436):1584–1594.

Fan, J. and Gijbels, I. (1996). *Local Polynomial Modelling and its Applications.* Chapman & Hall, London.

Fisher, R. A. (1935). Appendix to article by C. Bliss. *Annals of Applied Biology,* 22:164–165.

Fisher, R. A. (1936). The use of multiple measurements in taxonomic problems. *Annals of Eugenics,* 7:179–188.

Fisher, R. A. (1938). The statistical utilisation of multiple measurements. *Annals of Eugenics,* 8:376–386.

Fix, E. and Hodges, J. L. (1951). Discriminatory analysis, nonparametric discimination, consistency properties. Technical Report 4, United States Air Force, School of Aviation Medicine, Randolph Field, TX.

Freund, Y. and Schapire, R. E. (1996). Experiments with a new boosting algorithm. In *Machine Learning: Proc. Thirteenth International Conference,* pages 148–156. Morgan Kaufman.

Friedman, J. H. (1994). Flexible metric nearest neighbor classification. Technical report, Standford University.

Friedman, J. H. (2001). Greedy function approximation: A gradient boosting machine. *Annals of Statistics,* 29:1189–1232.

Friedman, J. H., Hastie, T., and Tibshirani, R. (2000). Additive logistic regression: A statistical view of boosting. *Annals of Statistics,* 28:337–407.

Gorman, R. P. and Sejnowski, T. J. (1988). Analysis of hidden units in a layered network trained to classify sonar targets. *Neural Networks,* 1:75–89.

Green, P. J. and Silverman, B. W. (1994). *Nonparametric Regression and Generalized Linear Models.* Chapman & Hall, London.

Hamerle, A. and Tutz, G. (1989). *Diskrete Modelle zur Analyse von Verweildauer und Lebenszeiten.* Campus, Frankfurt.

Hand, D. J. and Vinciotti, V. (2003). Local versus global models for classification problems: Fitting models where it matters. *The American Statistician,* 57(2):124–131.

Hastie, T. and Tibshirani, R. (1986). Generalized additive models. *Statistical Science,* 1:295–318.

Hastie, T. and Tibshirani, R. (1993). Varying-coefficient models. *Journal of the Royal Statistical Society*, B 55(4):757–796.

Hastie, T. and Tibshirani, R. (1996). Discriminant adaptive nearest neighbor classification. *IEEE Transactions on Pattern Analysis and Machine Intelligence*, 18(6):607–615.

Hastie, T. and Tibshirani, R. (2000). Bayesian backfitting. *Statistical Science*, 15(3):196–223.

Hastie, T., Tibshirani, R., and Friedman, J. (2001). *The Elements of Statistical Learning*. Springer, New York.

Hastie, T. J. and Tibshirani, R. J. (1990). *Generalized Additive Models*. Chapman & Hall, London.

Hoerl, A. E. and Kennard, R. W. (1970). Ridge regression: Biased estimation for nonorthogonal problems. *Technometrics*, 12(1):55–67.

Kalbfleisch, J. D. and Prentice, R. L. (2002). *The Statistical Analysis of Failure Time Data*. Wiley, Hoboken, New Jersey, 2nd edition.

Kauermann, G. and Tutz, G. (2000). Local likelihood estimates and bias reduction in varying coefficients models. *Journal of Nonparametric Statistics*, 12:343–371.

Klinger, A., Dannegger, F., and Ulm, K. (2000). Identifying and modelling prognostic factors with censored data. *Statistics in Medicine*, 19:601–615.

Lawless, J. F. (2002). *Statistical Models and Methods for Lifetime Data*. Wiley, Hoboken, New Jersey, 2nd edition.

Le Cessie, S. and van Houwelingen, J. C. (1992). Ridge estimators in logistic regression. *Applied Statistics*, 41(1):191–201.

Liu, H. and Motoda, H. (1998). *Feature Selection for Knowledge Discovery and Data Mining*. Kluwer, Boston.

Loader, C. (1999). *Local Regression and Likelihood*. Springer, New York.

Mallows, C. L. (1973). Some comments on C_p. *Technometrics*, 15(457):661–675.

Marx, B. D. and Eilers, P. H. C. (1998). Direct generalized additive modelling with penalized likelihood. *Computational Statistics and Data Analysis*, 28:193–209.

McCullagh, P. and Nelder, J. A. (1989). *Generalized Linear Models*. Chapman & Hall, 2nd edition.

Michie, D., Spiegelhalter, D. J., and Taylor, C. C. (1994). *Machine Learning, Neural and Statistical Classification*. Ellis Horwood, New York.

Nelder, J. and Wedderburn, R. W. M. (1972). Generalized linear models. *Journal of the Royal Statistical Society A*, 135:370–384.

R Development Core Team (2005). *R: A language and environment for statistical computing*. R Foundation for Statistical Computing, Vienna, Austria. ISBN 3-900051-07-0.

Ripley, B. D. (1996). *Pattern Recognition and Neural Networks*. Cambridge University Press, Cambridge.

Schaal, S., Vijayakumar, S., and Atkeson, C. G. (1998). Local dimensionality reduction. In Jordan, M. I., Kearns, M. J., and Solla, S. A., editors, *Advances in Neural Information Processing Systems 10*. MIT Press, Cambridge, MA.

Schapire, R. E. (1990). The strength of weak learnability. *Machine Learning*, 5:197–227.

Thompson, W. A. (1977). On the treatment of grouped observations in life studies. *Biometrics*, 33:463–470.

Tibshirani, R. (1996). Regression shrinkage and selection via the lasso. *Journal of the Royal Statistical Society*, B 58(1):267–288.

Tutz, G. and Binder, H. (2004a). Flexible modelling of discrete failure time including time-varying smooth effects. *Statistics in Medicine*, 23(15):2445–2461.

Tutz, G. and Binder, H. (2004b). Generalized additive modelling with implicit variable selection by likelihood based boosting. Discussion Paper 401, SFB 386, Ludwig-Maximilians-University Munich.

Tutz, G. and Binder, H. (2005a). Boosting ridge regression. Discussion Paper 418, SFB 386, Ludwig-Maximilians-University Munich.

Tutz, G. and Binder, H. (2005b). Localized classification. *Statistics and Computing*, 15(3):155–166.

Tutz, G. and Scholz, T. (2004). Semiparametric modelling of multicategorical data. *Journal of Statistical Computation and Simulation*, 74(3):183–200.

Venables, W. N. and Ripley, B. D. (1999). *Modern Applied Statistics With S-Plus*. Springer, 3rd edition.

Wood, S. (2004). Stable and efficient multiple smoothing parameter estimation for generalized additive models. *Journal of the American Statistical Association*, 99(467):673–686.

Wood, S. N. (2000). Modelling and smoothing parameter estimation with multiple quadratic penalties. *Jounal of the Royal Statistical Society B*, 62(2):413–428.

Zou, H. and Hastie, T. (2005). Regularization and variable selection via the elastic net. *Journal of the Royal Statistical Society B*, 67(2):301–320.

Lebenslauf

Persönliche Daten

Name:	Harald Binder
Geburtsdatum/-ort:	19. Juli 1976 in Ingolstadt
Staatsangehörigkeit:	deutsch

Ausbildung/Studium

09/1982 – 07/1986	Grundschule Manching
09/1986 – 06/1995	Apian Gymnasium Ingolstadt
06/1995	Abitur
08/1995 – 08/1996	Zivildienst
09/1996 – 09/2000	Diplomstudiengang Psychologie an der Universität Regensburg
05/1998	Vordiplom in Psychologie, Universität Regensburg
09/2000	Diplom in Psychologie, Universität Regensburg
09/2000 – 06/2001	"Mathematical Behavioral Science"-Programm an der University of California, Irvine, USA
04/2001	Master of Arts in Social Science, University of California, Irvine
seit 07/2001	Promotion in Statistik bei Prof. Dr. Gerhard Tutz an der Ludwig-Maximilians-Universität München

Beruflicher Werdegang

04/1997 – 09/2000	Studentische Hilfskraft an der Universität Regensburg
09/2000 – 12/2000	Research Assistant, University of California, Irvine
01/2001 – 06/2001	Teaching Assistant, University of California, Irvine
seit 07/2001	Wissenschaftlicher Mitarbeiter an der Klinik und Poliklinik für Psychiatrie, Psychosomatik und Psychotherapie der Universität Regensburg